The Ageing Immune System and Health

Valquiria Bueno • Janet M. Lord
Thomas A. Jackson
Editors

The Ageing Immune System and Health

Springer

Editors
Valquiria Bueno
Department of Microbiology,
  Immunology and Parasitology
UNIFESP Federal University of São Paulo
São Paulo, São Paulo, Brazil

Thomas A. Jackson
Institute of Inflammation and Ageing
University of Birmingham
University of Birmingham Research
  Laboratories
Birmingham, UK

Queen Elizabeth Hospital
University Hospitals Birmingham
  Foundation NHS Trust
Edgbaston, Birmingham UK

Janet M. Lord
MRC-ARUK Centre for Musculoskeletal
  Ageing Research
Institute of Inflammation and Ageing
Birmingham University Medical School
  Birmingham
Birmingham, UK

ISBN 978-3-319-82803-9   ISBN 978-3-319-43365-3   (eBook)
DOI 10.1007/978-3-319-43365-3

Printed on acid-free paper

This Springer imprint is published by Springer Nature
The registered company is Springer International Publishing AG
The registered company address is: Gewerbestrasse 11, 6330 Cham, Switzerland

Preface

The world population is undergoing a major demographic shift. As birth rates decline and people live longer due to advances in public health and medical science, the proportion of people over 65 is increasing. Moreover, people over 85 are the fastest growing section of society and often referred to as the "oldest old". However, while life expectancy is rising at a rate of approximately 2 years per decade, this is not accompanied with a similar increase in disability-free life expectancy. So people are living longer, but in poorer health. However, as we age, we observe a greater heterogeneity of ability and health. The variation in, say, walking speed is far greater in a group of 70 year olds, than in a group on 20 year olds. This makes the study of ageing and the factors driving that heterogeneity of health and functional ability in old age vital.

Infectious disease in older people is associated with greater morbidity and mortality, so it is plausible that age-related changes to the immune system are central to this. The influenza virus is associated with greater hospitalisation and mortality in older people, and older people are more susceptible to complications of infection such as delirium. The study of the immune system across the lifespan has demonstrated that as we age the immune system undergoes a decline in function, termed immunosenescence. However, as discussed in Chaps. 1 and 2 the decline in function is not universal across all aspects of the immune system, and neither is the magnitude of functional loss similar between individuals. The theory of inflammageing, which represents a chronic low grade inflammatory state in older people, has been described as a major consequence of immunosenescence, though lifestyle factors such as reduced physical activity and increased adiposity also play a major role. Importantly, inflammageing may well explain the greater burden of disease in older people as increased systemic inflammation has been associated with greater risk of most of the age-related conditions including cardiovascular disease, cancer, sarcopaenia and dementia.

In poor health, older people accumulate disease, described as multimorbidity. This in turn means traditional single system based health care becomes less valid as each system affected by disease impacts on other systems. This leads some older people to be at greater risk of adverse events such as disability and death. The syndrome of this increased vulnerability is described as frailty, and increasing fundamental evidence is emerging that suggests immunosenescence and inflammageing may underpin frailty and this is discussed in Chap. 9. Thus frailty is seen as one clinical manifestation of immunosenescence.

The understanding of how the immune system changes with age and can potentially be manipulated will impact on our current knowledge of older people living healthy lives and also can direct actions on how to improve the care of older people in ill health. The major aim is to improve disability-free life expectancy, so thereby adding life to years. The role that increasing physical activity may play in reducing immunosenescence and ill health in old age is covered in Chap. 10.

Despite the importance of the ageing immune system on both good and poor health in older people, there is a dearth of textbooks bringing together all aspects of the topic. In this book we aim to present up-to-date reviews on the key topics in the understanding of Ageing, Immune System, and Health. It is aimed at fundamental scientists and clinicians with an interest in ageing or the immune system. Each chapter aims to highlight current knowledge and also highlight knowledge gaps to stimulate further research.

Chapters include state-of-the-art reviews on immunosenescence in both the innate and adaptive immune systems. Others follow ageing and immunity in specific systems; lung senescence and epigenetics, as well as specific disease processes; cancer and cytomegalovirus infection. Vaccination is discussed in relation to older people, and the clinical and fundamental aspects of frailty are discussed.

Our goal is that improved recognition of the role played by a compromised immune system in ill health in old age, combined with the understanding that this can to a large extent be attenuated by lifestyle choices, will result in public health policy that ensures old age is enjoyed and not endured!

São Paulo, Brazil   Valquiria Bueno
Birmingham, UK   Thomas A. Jackson
Birmingham, UK   Janet M. Lord

Acknowledgments

The book *The Ageing Immune System and Health* is part of a UK-Brazil collaborative project with financial support to Prof. Bueno and Prof. Lord from FAPESP – São Paulo Research Foundation (2014/50261-8). We also acknowledge the support of the MRC-Arthritis Research UK Centre for Musculoskeletal Ageing Research to Prof. Lord.

Contents

Contributors

Ayodeji A. Asuni Department Neurodegeneration in vivo, Valby, Denmark

David B. Bartlett Duke Molecular Physiology Unit, Duke School of Medicine, Duke University, Durham, NC, USA

Duke Molecular Physiology Institute, Department of Medicine, Duke University Medical Center, Durham, NC, USA

Valquiria Bueno Department of Microbiology, Immunology and Parasitology, UNIFESP Federal University of São Paulo, São Paulo, Brazil

Nora Manoukian Forones Department of Medicine, UNIFESP Federal University of São Paulo, São Paulo, Brazil

Sarah J. Goodman Center for Molecular Medicine and Therapeutics, University of British Columbia, BC Children's Hospital, Vancouver, BC, Canada

Carolyn A. Greig MRC-Arthritis Research UK Centre for Musculoskeletal Ageing Research, University of Birmingham, Edgbaston, Birmingham, UK

School of Sport, Exercise and Rehabilitation Sciences, University of Birmingham, Edgbaston, Birmingham, UK

Klaus Hamprecht Institute of Medical Virology, University of Tübingen, Tübingen, Germany

Jon Hazeldine MRC-ARUK Centre for Musculoskeletal Ageing Research, Institute of Inflammation and Ageing, Birmingham University Medical School, Birmingham, UK

Sian M. Henson Centre for Microvascular Research, William Harvey Research Institute, Barts and The London School of Medicine and Dentistry, Queen Mary, University of London, London, UK

Kim M. Huffman Duke Molecular Physiology Unit, Duke School of Medicine, Duke University, Durham, NC, USA

Thomas A. Jackson Institute of Inflammation and Ageing, University of Birmingham, University of Birmingham Research Laboratories, Queen Elizabeth Hospital, Birmingham, UK

Queen Elizabeth Hospital, University Hospitals Birmingham Foundation NHS Trust, Edgbaston, Birmingham, UK

Meaghan J. Jones Center for Molecular Medicine and Therapeutics, The University of British Columbia, BC Children's Hospital, Vancouver, BC, Canada

Michael S. Kobor Center for Molecular Medicine and Therapeutics, The University of British Columbia, BC Children's Hospital, Vancouver, BC, Canada

Krisztian Kvell Department of Pharmaceutical Biotechnology, School of Pharmacy, and Szentagothai Research Center, University of Pecs, Pecs, Hungary

Janet M. Lord MRC-ARUK Centre for Musculoskeletal Ageing Research, Institute of Inflammation and Ageing, Birmingham University Medical School, Birmingham, UK

Lisa M. McEwen Center for Molecular Medicine and Therapeutics, The University of British Columbia, BC Children's Hospital, Vancouver, BC, Canada

Ludmila Müller Max Planck Institute for Human Development, Berlin, Germany

Graham Pawelec Center for Medical Research, University of Tübingen, Tübingen, Germany

Division of Cancer Studies, Faculty of Life Sciences and Medicine, King's College London, London, UK

The John van Geest Cancer Research Centre, School of Science and Technology, Nottingham Trent University, Nottingham, UK

Judit E. Pongracz Department of Pharmaceutical Biotechnology, School of Pharmacy, and Szentagothai Research Center, University of Pecs, Pecs, Hungary

Mohammad Ahsan Tariq MRC-ARUK Centre for Musculoskeletal Ageing Research, Institute of Inflammation and Ageing, Birmingham University Medical School, Birmingham, UK

Jessica L. Teeling Centre for Biological Sciences, University of Southampton, Southampton, UK

Birgit Weinberger Research Institute for Biomedical Aging Research, University of Innsbruck, Innsbruck, Austria

Daisy Wilson Institute of Inflammation and Ageing, University of Birmingham, University of Birmingham Research Laboratories, Queen Elizabeth Hospital, Edgbaston, Birmingham, UK

MRC-Arthritis Research UK Centre for Musculoskeletal Ageing Research, University of Birmingham, Edgbaston, Birmingham, UK

Innate Immunosenescence and Its Impact on Health in Old Age

1

Mohammad Ahsan Tariq, Jon Hazeldine, and Janet M. Lord

Abstract

Physiological ageing is associated with significant re-modelling of the immune system. Termed immunosenescence, age-related changes have been described in the composition, phenotype and function of both the innate and adaptive arms of the immune system. As the first line of defence against invading pathogens, age-associated alterations in innate immunity have been linked to the increased infection-related morbidity and mortality rates reported by older adults. However, is the only consequence of innate immunosenescence an increased susceptibility to infection? With data emerging demonstrating a role for innate immune cells in other biological processes besides host protection, such as wound healing, clearance of senescent cells and the resolution of inflammation, it is likely that innate immunosenescence has more far reaching consequences for the health and well-being of older adults than originally thought. Here, we provide an overview of the alterations that occur in innate immunity with age, highlighting studies that have uncovered the molecular mechanisms that underlie the changes they describe. Furthermore, we discuss the possible implications of innate immunosenescence for older adults beyond the much-discussed increased incidence and severity of infection.

Keywords

Innate immunity • Ageing • Neutrophil • NK cell • Monocyte • Inflammageing

M.A. Tariq • J. Hazeldine • J.M. Lord (✉)
MRC-ARUK Centre for Musculoskeletal Ageing Research, Institute of Inflammation and Ageing, Birmingham University Medical School, Birmingham B15 2TT, UK
e-mail: J.M.Lord@bham.ac.uk

© Springer International Publishing Switzerland 2017
V. Bueno et al. (eds.), *The Ageing Immune System and Health*,
DOI 10.1007/978-3-319-43365-3_1

1.1   Introduction

Recent forecasts estimate that the proportion of individuals aged ≥60 years, who accounted for 10 % of the global population in the year 2000, will constitute 21 % of the world's population by 2050 [1]. A testament to advancements in medical care and public health policies, an increase in life expectancy should be viewed favourably. However, with ageing defined as "*the increasing frailty of an organism with time that reduces their ability to deal with stress, resulting in an increased chance of disease*" [2], an extension in life expectancy does not necessarily mean these "additional years" will be experienced in good health.

Infectious diseases are a common occurrence amongst older people, with this section of society reporting increased episodes of urinary tract infections [3], influenza [4] and community-acquired pneumonia [5] when compared to their younger counterparts. In addition, infection severity is greater, with older adults reporting significantly increased infection-related morbidity and mortality rates [6, 7]. Thought to underlie this increased incidence and severity of infection are age-associated changes in immune function, a phenomenon termed *immunosenescence*. Whilst it has been known for many years that ageing has a profound effect upon adaptive immunity [8], only recently has it become evident that the innate arm of the immune system undergoes considerable age-related re-modelling. For instance, alongside alterations in the efficiency of innate barriers and the composition of the humoral arm of innate immunity [9, 10], significant differences exist in the composition, phenotype and function of the cellular arm of innate immunity between young and older adults. In this chapter, we provide a detailed overview of the impact that age has on four cell types central to the innate immune response; neutrophils, natural killer (NK) cells, dendritic cells (DCs) and monocytes/macrophages. Furthermore, with it now recognised that the function of the innate immune system extends beyond the recognition and elimination of pathogens [11–14], we discuss how innate immunosenescence may have more far reaching consequences for the health of older adults than simply increasing their susceptibility to infection.

1.2   Ageing of the Innate Immune System

1.2.1   Neutrophils

Neutrophils are the most abundant leukocyte in circulation and provide immediate frontline protection against rapidly dividing bacteria, fungi and yeast. The first step in neutrophil anti-microbial defence is their extravasation from the bloodstream and migration to the site of infection. Whilst age appears to have no effect upon the speed at which neutrophils migrate towards chemotactic signals in vitro [15], the directional accuracy of neutrophil migration to inflammatory agonists (e.g. interleukin-8 (IL-8, CXCL8)) as well as bacterial peptides (e.g. formyl-methionine-leucine-phenylalanine (fMLP)) is significantly reduced [15]. Recently, we attributed this age-related aberration in neutrophil   chemotaxis to constitutive activation of phosphoinositide 3-kinase (PI3-K), a lipid kinase implicated in cell polarisation and propulsion via its generation of phosphoinositol 3,4,5-triphosphate (PIP$_3$) at the leading edge of the cell [15]. Selective inhibition of the PI3-K isoforms PI3-K δ and γ in neutrophils from older adults significantly improved their migratory accuracy towards IL-8 [15]. Conversely, increasing PIP$_3$ levels in neutrophils from young donors, resulted in reduced chemotaxis similar to that seen in older donors [15]. Given that in a longitudinal study of older adults, reduced neutrophil chemotaxis at baseline was associated with "non-survival" following infectious episodes [16], our work on PI3-K suggests that therapeutic targeting of this kinase could be a novel approach by which to improve patient outcome in times of infection [15]. Whilst age-related impairments in neutrophil migration have been observed in vitro [15, 17], conflicting data has emerged from in vivo studies, where neutrophil chemotaxis has been reported to be either unaffected with age [18] or significantly reduced [19, 20]. In the most recent of these studies, Nacionales and colleagues recovered significantly fewer neutrophils from the bronchoalveolar lavage fluid of aged mice in a model of pneumonia infection [20]. Reduced expression of adhesion molecules on the surface of activated endothelium could be one mechanism that explains the aberrant migration of neutrophils in aged mice [19].

Upon arrival at sites of infection, neutrophils attempt to contain invading pathogens via phagocytosis. To date, no age-related impairments have been reported in the ability of neutrophils to engulf non-opsonised particles [21]. However, neutrophils from older adults exhibit significantly reduced phagocytic activity towards complement and immunoglobulin (Ig)-opsonised pathogens [22, 23]. We have attributed the age-related impairment in neutrophil uptake of Ig-coated microbes to reduced expression of CD16, an Fc receptor, whose surface density positively correlates with the uptake of Ig-opsonised pathogens [22]. In contrast, the surface expression of complement receptors are comparable between neutrophils from young and old donors [24], suggesting intrinsic signalling defects are likely to be responsible for reduced phagocytic activity of neutrophils from older adults towards complement-coated pathogens. Once phagocytosed, pathogens are exposed to a harsh microbicidal environment, the creation of which requires the generation of reactive oxygen species (ROS). Currently, the effect that physiological ageing has on neutrophil ROS generation is unclear, with study outcomes seemingly dependent upon the type of activating stimulus. For example, no age-associated differences have been described for ROS generation initiated by *Candida albicans* or *Escherichia coli* (*E. coli*) stimulation [23, 25], whereas neutrophils from older adults generate significantly fewer ROS when challenged with *Staphylococcus aureus* (*S. aureus*), fMLP or granulocyte macrophage-colony stimulating factor (GM-CSF) [17, 23, 25].

In a seminal paper published in 2004, Brinkmann and co-workers demonstrated that following microbial or pro-inflammatory cytokine stimulation, neutrophils release into the extracellular environment their nuclear DNA [26]. Termed neutrophil extracellular traps (NETs), these structures are decorated with an array of granule and cytosol-derived peptides and proteases and have been shown to capture and in some cases disarm gram positive and negative bacteria [26]. To our knowledge, only two studies have investigated whether NET formation is altered during "healthy" ageing. In response to stimulation with the diacylglycerol mimetic phorbol 12-myristate 13-acetate (PMA) or, methicillin-resistant *S. aureus* (MRSA), Tseng *et al.* found in vitro NET production by neutrophils from aged mice was significantly lower than that of neutrophils from young mice, a defect they also observed in an in vivo model of MRSA infection [27]. In agreement with these findings, we showed that human ageing also results in impaired NET formation [28]. Compared to those isolated from their older counterparts, we found tumour necrosis factor-alpha (TNF-α) primed neutrophils from younger adults generated significantly more NETs following IL-8 or lipopolysaccharide stimulation [28]. Mechanistically, this aberration in NET formation was attributed to an age-related decline in ROS generation, a crucial prerequisite for NET production [28, 29].

Aside from the controversy that exists regarding the impact of age on ROS production [17, 23, 25], neutrophils from older adults clearly exhibit defects in several key defensive mechanisms, namely chemotaxis [15, 19, 20], phagocytosis of opsonised pathogens [22, 23] and NET formation [27, 28]. Given this near global impairment in neutrophil function, alterations to a generic signalling element rather than defects in molecules specific to each anti-microbial defence strategy is likely to explain the aberrations in neutrophil function that occur with age. In support of this idea, ageing in rodents is associated with a significant increase in neutrophil membrane fluidity, which coincides with a marked reduction in neutrophil function [30]. Moreover, human studies have shown the recruitment of receptors into lipid rafts, specialised microdomains enriched for phospholipids and cholesterol that assist in signal transduction, is markedly reduced with age [17, 31]. That proximal signalling defects contribute at least in part to the impaired functions of aged neutrophils is further demonstrated by the fact that no age-related differences in ROS production [17, 28, 30] or NET formation [28] have been reported for neutrophils treated with PMA, a stimulus that activates cells independently of surface membrane receptors.

1.2.2   Natural Killer Cells (NK)

Involved in the direct recognition and elimination of virus-infected, stressed and malignant cells, NK cells are large granular lymphocytes that comprise approximately 10–15 % of the circulating lymphocyte pool. In humans, NK cells are defined by a CD3$^-$56$^+$ surface phenotype. However, they are not a homogeneous population as based on the differential surface expression of CD56, NK cells are assigned to one of two major subsets, namely CD56^{DIM} and CD56^{BRIGHT}. Despite the fact that ageing results in a reduction in NK cell production and proliferation [32], numerous studies have reported a significant increase in the percentage and/or absolute number of CD3$^-$56$^+$ cells with age [33, 34], suggesting the presence of long-lived NK cells in the circulation of older adults [32]. At the subset level, ageing is associated with an increased proportion of CD56^{DIM} NK cells and a reduction in the proportion of CD56^{BRIGHT} NK cells [33, 34], alterations that culminate in a significantly increased circulating CD56^{DIM}:CD56^{BRIGHT} ratio with age [33, 35].

Results of two recent murine models of infection suggest ageing results in aberrant NK cell migration. Following influenza [36] or mousepox [37] infection, migration of NK cells to the spleen, lungs or lymph nodes was shown to be significantly reduced in aged mice during the early post infection phase, a defect that was assigned to an age-related impairment in NK cell maturation [37]. Whilst no study to date has investigated the effect of age on the migratory capacity of human NK cells, groups have examined its impact on the expression of adhesion and chemokine receptors. With the exception of the IL-8 receptor CXCR1, whose surface density is reduced with age [38], comparable expression of the adhesion molecule CD2 and the chemokine receptors CCR3 and CCR5 has been reported for NK cells from young and older adults [35, 38].

Granule exocytosis and death receptor ligation are two contact-dependent mechanisms utilised by NK cells to eliminate transformed cells [39]. Of these, granule exocytosis, which is characterised by the secretion of the pore-forming protein perforin and the serine protease granzyme B, is the primary cytotoxic mechanism of NK cells. Numerous studies have examined the impact of age on this form of NK cell defence [40], with the general consensus that at the single cell level, NK cell cytotoxicity (NKCC) is reduced with age [33, 41, 42]. Given that NK cells from older adults recognise and bind to transformed cells as efficiently as those from younger subjects [42, 43], a post-binding defect is thought to underlie the age-related impairment in NKCC. At the molecular level, the induction of NKCC is regulated by signals transmitted through surface expressed activatory and inhibitory receptors. Currently, it is unclear as to what effect ageing has on NK cell receptor expression, with a series of conflicting observations reported by a number of independent groups (reviewed in [40]). However, data has been presented that suggests defective intracellular signalling may underlie the age-related impairment in NKCC. Following target cell stimulation, Mariani *et al.* found that due to an age-related delay in phosphatidylinositol 4,5-bisphosphate hydrolysis, NK cells from older adults generated significantly lower amounts of the second messenger inositol 1,4,5-triphosphate [43]. If responsible for the age-associated reduction in NKCC, one would predict a functional consequence of this impaired intracellular signalling. Recently, we demonstrated that when bound to transformed cells, NK cells from older adults exhibit impaired polarisation of perforin-containing lytic granules to the immunological synapse, a defect that was associated with reduced perforin secretion [33]. Importantly, we observed no age-related difference in perforin release when NK cells were stimulated with PMA and ionomycin (J. Hazeldine, unpublished observations). As agents that bypass cell surface receptors, this comparable functional response of NK cells from young and older adults to PMA and ionomycin treatment suggests that it is defective signalling proximal to the NK cell membrane that underlies the age-related impairment in NKCC.

Although renowned for their cytotoxic nature, it is becoming increasingly recognised that NK cells are a rich early source of immunoregulatory cytokines and chemokines. For example, activated NK cells secrete TNF-α, IL-8 and interferon gamma (IFN-γ), bestowing upon them the ability to amplify on-going innate immune responses and influence the early phases of adaptive immune responses

[44–46]. When challenged with inflammatory cytokines, NK cells from older adults respond with significantly increased secretion of IFN-γ, macrophage inflammatory protein-1-alpha (MIP-1α) and IL-8. However, the levels generated are significantly lower than those produced by NK cells from younger subjects [45, 46]. Akin to humans, NK cells from aged mice exhibit aberrant cytokine production, with significant impairments reported in IFN-γ production in models of influenza infection [36] and lipopolysaccharide challenge [47].

The recurrent and severe episodes of viral infection experienced by subjects with selective NK cell deficiency demonstrate the importance of NK cells in host defence [48]. However, do the more subtle abovementioned age-related changes in NK cell function impact upon the health and/or well-being of older adults? To date, very few studies have attempted to answer this question. However, from the data available, reduced NKCC does appear to have clinical consequences. For example, retrospective and prospective studies have reported relationships between low NK cell activity in older adults and (1) a past history of severe infection, (2) an increased risk of future infection, (3) a reduced probability of surviving infectious episodes and (4) infectious morbidity [49–51]. Related to this increased risk of infection, reduced NKCC prior to and following influenza vaccination in older adults has been shown to be associated with reduced protective anti-hemagglutinin titres, worsened health status and an increased incidence of respiratory tract infection [52].

1.2.3 Monocytes/Macrophages

Comprising approximately 5–10 % of the circulating leukocyte pool, monocytes are a heterogeneous population of immune cells, which based on the differential surface expression of CD14 and the Fc receptor CD16 are categorised into one of three distinct subsets, namely classical (CD14^{++}16^{-}), intermediate (CD14^{++}16^{+}) and non-classical (CD14^{+}16^{++}) [53]. Whilst age has no effect upon the frequency or absolute number of monocytes [54, 55], the composition of the monocyte pool is markedly different in older adults, who present with an increased frequency of non-classical and intermediate monocytes, and fewer classical monocytes when compared to their younger counterparts [55–57].

Providing frontline protection against viruses, bacteria and transformed cells, monocytes and macrophages are equipped with an array of anti-microbial mechanisms, which include chemotaxis, ROS generation, phagocytosis and antigen presentation. The results of rodent and human-based studies that have examined the impact of age on these functions are summarised in Fig. 1.1. In terms of macrophage function, conflicting observations are often reported, the result primarily of inter-study differences in such aspects of experimental design as the type of mouse strain, the source of tissue macrophages and the definition of "young" and "aged" mice.

The extent to which ageing impacts upon cytokine/chemokine generation by monocytes/macrophages has been the subject of considerable interest from immune gerontologists. Pathogen recognition by monocytes and macrophages is achieved primarily through their expression of toll like receptors (TLRs), a family of endosomal

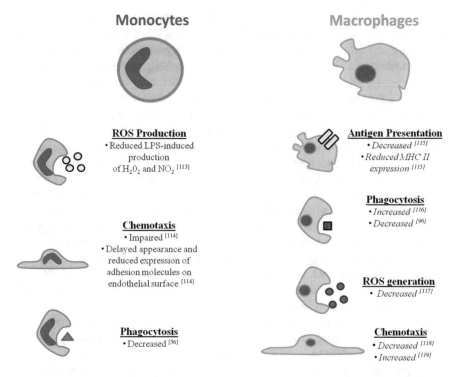

Fig. 1.1 Summary of age-related changes in monocyte and macrophage function. *Italic font* indicates results obtained from rodent-based studies

and surface residing receptors that recognise structurally conserved molecules shared by viruses, bacteria and fungi [58]. In the most extensive study to date that have investigated the effect of age on TLR-induced cytokine production, van Duin *et al.* measured intracellular IL-6 and TNF-α levels in monocytes isolated from young and older adults following TLR1/2, TLR2/6, TLR4, TLR5, TLR7 and/or TLR8 stimulation [59]. After controlling for a range of potential confounding factors such as medication, gender and vaccination history, the group reported a significant age-related reduction in TLR1/2-induced IL-6 and TNF-α synthesis [59]. Subsequent studies have confirmed this observation and demonstrated the aberration in TLR1/2-induced IL-6 production is a feature of all monocyte subsets [57]. In contrast to the impaired response to TLR1/2 stimulation, monocytes from older adults generate significantly greater amounts of TNF-α and IL-8 upon TLR4 or TLR5 stimulation respectively when compared to those from their younger counterparts [54, 56]. Whilst heightened responses of intermediate and non-classical monocytes drive the age-related enhancement in TLR4-induced TNF-α production [56], it is currently unknown which monocyte subset(s) is responsible for the increase in IL-8 synthesis following TLR5 stimulation. However, reports of an age-associated increase in TLR5 expression and signalling capacity provides a potential mechanistic explanation for the enhanced functional response of monocytes from older adults to TLR5 stimulation [54].

Based on the abovementioned age-related elevation in TLR5-induced cytokine production, immunologists are currently investigating whether manipulation of this signalling pathway could enhance the immune response elicited by older adults in times of infectious challenge. Early results are promising, with studies demonstrating a potential role for flagellin, the ligand for TLR5, as a vaccine adjuvant. In a model of Streptococcus pneumonaie infection, Lim and co-workers reported a 60 % increase in the survival rates of aged mice that received a combined pneumococcal surface protein A (PspA)-flagellin vaccine when compared to those mice vaccinated with PspA alone [60]. This increased survival rate was associated with significantly elevated levels of PspA-specific antibodies, suggesting that through activation of TLR5, robust immune responses can be generated in the aged host [60]. A similar approach has also proven successful in older adults, with a recombinant hemagglutinin influenza-flagellin fusion vaccine inducing seroconversion and seroprotection rates of 75 % and 98 % respectively in adults aged $\geq$65 years [61]. Interestingly, monocyte responsiveness in vitro to TLR stimulation is highly associated with vaccination responses. In a cohort of 162 young (21–30 years) and older ($\geq$65 years) subjects, an age-related reduction in TLR-induced up-regulation of the co-stimulatory molecule CD80 on the surface of monocytes was reported [62]. As subsequent analysis revealed an increased frequency of CD80$^+$ monocytes post TLR stimulation was associated with an increased antibody response to influenza vaccination, prior assessment of monocyte reactivity to TLR stimulation may identify older adults unlikely to respond to vaccination [62].

To gain an insight into the mechanism(s) underlying the age-associated reduction in TLR-induced cytokine production, several groups have investigated the impact of age on the expression and/or signalling capacity of surface residing TLRs. Whilst TLR2 expression is comparable on monocytes from young and older adults [54, 57, 59], a significant age-related decline in TLR1 and TLR4 expression has been reported, which in the case of TLR1 is associated with reduced phosphorylation of the mitogen activated protein kinase (MAPK), extracellular signal-regulated kinase 1/2 (ERK 1/2) post TLR1/2 stimulation [57, 59]. On this theme of intracellular signalling, Olivieri et al. recently outlined how microRNAs (miRNAs) may be contributing to the dysregulation in TLR-induced cytokine production with age [63]. A family of short single-stranded RNA molecules that regulate gene expression, age-related changes in the levels of specific miRNAs may influence TLR-induced signalling by modulating the activity and/or expression of adaptor molecules and kinases central to TLR signal transduction [63].

Akin to monocytes, TLR-induced cytokine production by macrophages is significantly reduced with age. Following TLR1/2, TLR2/6, TLR3, TLR4, TLR5 and/or TLR9 stimulation, splenic, alveolar and peritoneal macrophages from aged mice secrete significantly less TNF-α, IL-6, IL-1β or IL-12 when compared to macrophages from their younger counterparts [64–68]. Mechanistically, this age-related aberration in TLR-induced pro-inflammatory cytokine production has been assigned to reduced expression of TLRs, increased production of the anti-inflammatory cytokine IL-10, altered expression of miRNAs and defective MAPK and protein kinase C signalling [64–69]. In one of the few studies investigating the impact of age on human macrophage function, Agius and colleagues assessed TNF-α production by skin macrophages in a model of cutaneous delayed type hypersensitivity and reported a significantly reduced percentage of TNF-α⁺ macrophages in skin biopsies taken from older adults [70]. Interestingly, rather than a defect intrinsic to macrophages, the group attributed this reduced production of TNF-α to an increased proportion in the skin of older adults of regulatory T cells, an immune cell subset that suppresses TNF-α production by macrophages [70]. Indeed, when macrophages isolated from the skin of young and older adults were subjected to TLR stimulation in vitro, no age-related differences in TNF-α production were observed [70].

1.2.4 Dendritic Cells (DCs)

Residing at the interface of innate and adaptive immunity, DCs are a heterogeneous family of circulating and tissue resident immune cells that based on their cellular origins are assigned to one of two distinct subsets. Myeloid-derived DCs (mDCs), which include Langerhans cells (LCs) in the skin and conventional DCs (cDCs) in the blood assist, in the case of LCs, in the initiation of adaptive immune responses whereas DCs of lymphoid descent, which are represented by blood-borne plasmacytoid DCs (pDCs) are involved in anti-viral immune responses via their secretion of IFN-α.

Human-based immunogerontological studies that have investigated the impact of age on DC number have focused primarily upon those subsets present in peripheral blood, namely cDCs and pDCs. Whilst age appears to have no effect upon the frequency of cDCs [71–73], its impact on pDCs is unclear, with some studies revealing a marked reduction in the percentage and/or absolute number of this DC subset with age [72–74], and others reporting no difference between young and older adults [71, 75]. In respect of tissue-resident DCs, a significant age-associated reduction in the frequency of LCs has been described [76].

1.2.4.1 pDCs

Several groups have demonstrated that following engagement of TLR7, TLR8 or TLR9, pDCs from young individuals generate significantly more IFN-α, TNF-α and IL-6 than pDCs obtained from older adults [72, 73, 77, 78]. This age-related reduction in TLR-induced cytokine production has been attributed to a decline in TLR expression and aberrant intracellular signalling [78]. For example, although not observed by all groups [79], ageing appears to be associated with reduced expression of TLR7 and TLR9 [72, 73], whilst impaired up-regulation of PI3-K and the transcription factor IFN regulatory factor-7 has been observed in TLR9-stimulated pDCs from aged donors [80].

Through the production of IFN-α, pDCs augment anti-viral immune responses by enhancing the cytotoxic activity of NK and CD8⁺ T cells. Thus, an age-related impairment in IFN-α secretion may contribute to the increased incidence of viral infections amongst older adults. Supporting this theory, Panda and co-workers found TLR-induced production of IFN-α by pDCs was significantly associated with both seroconversion and seroprotection following influenza vaccination in a cohort of young (21–30 years) and older (≥65 years) adults [73], whilst Sridharan *et al.* demonstrated that via reduced production of IFN-α, pDCs from aged donors were unable to enhance the cytotoxicity of resting CD8⁺ T cells [79].

1.2.4.2 mDCs

Tissue-resident DCs continually survey and sample their local environment for invading microbes, whose recognition triggers DC maturation and their migration to secondary lymphoid organs. Both human and murine-based studies have demonstrated a significant impairment in DC migration with age [76, 81]. The age-associated aberration in the in vivo migration of murine DCs has been attributed to an altered microenvironment as well as impaired signalling from surface expressed chemokine receptors [81, 82].

Increasing experimental evidence suggests that the stimulatory capacity of mDCs wanes with age. Compared to those isolated from young controls, aged mDCs are less effective at inducing T cell proliferation [78, 81, 83], IFN-γ secretion [83, 84] and T cell cytotoxicity [81, 84], defects that would be expected to lead to a diminished T cell-mediated immune response. However, with several studies reporting that antigen presentation and expression of the co-stimulatory molecules CD80/86 are comparable between DCs following stimulation, it cannot be ruled out that the weakened T cell response of aged individuals may be the result of consequence of intrinsic T cell defects rather than impaired DC function [85].

Whether ageing is associated with alterations in pro-inflammatory cytokine production by mDCs is an area of controversy, with studies reporting increased [71], decreased [75, 86] or comparable [72] generation of TNF-α, IL-6 and IL-12 by aged mDCs in response to TLR stimulation when compared to that secreted by young mDCs. Studies that have reported a decline in cytokine production have attributed it to an age-related reduction in TLR expression [73], impaired intracellular signalling [87] and an age-associated increase in basal cytokine levels [73].

1.3   Innate Immunosenescence and Its Impact on Healthy Ageing

Early recognition and elimination of invading pathogens is the defining feature of innate immunity. However, it has become apparent in recent years that the function of the innate immune system extends beyond host protection, with studies demonstrating roles for innate immune cells in immune modulation [88], the resolution of inflammation [12, 13], wound healing [89] and the clearance of senescent cells [11]. Thus, with these observations in mind, could innate immunosenescence have more far reaching consequences for older adults than just simply increasing their susceptibility to infection? In this section, we consider the wider-impact that age-related changes in NK cell (Fig. 1.2), neutrophil and macrophage biology may have on older adults and discuss how a low-grade chronic up-regulation in circulating inflammatory mediators may underlie the development of several age-related pathologies.

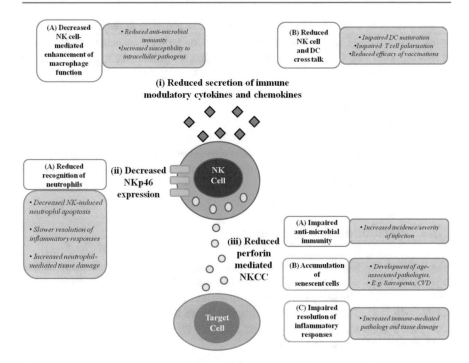

Fig. 1.2 NK cell immunosenescence and its impact upon the health and well-being of older adults. Aside from the associations reported in older adults between reduced NK cell function and increased infection-related morbidity and mortality rates [65–67], age-related changes in NK cell cytokine/chemokine production (**i**), surface phenotype (**ii**) and NKCC (**iii**) may have additional consequences for the health of older adults. These include; (a) the accumulation of senescent cells and the subsequent development of such age-related pathologies as sarcopenia and cardiovascular disease (CVD) [14], (b) impaired resolution of inflammatory responses and increased immune-mediated tissue damage as a consequence of impaired NK-mediated induction of neutrophil, DC and CD4+ T cell lysis and (c) reduced vaccine efficacy as a result of aberrant NK-DC cross talk

1.3.1   NK Cell-Mediated Clearance of Senescent Cells

A feature of physiological ageing is the accumulation of senescent cells. These cells, which have been detected in bone and skin samples obtained from older adults [90, 91], reside in a state of irreversible cell cycle arrest, yet remain viable and meta-bolically active. Via their secretion of growth factors, pro-inflammatory cytokines, and proteases, senescent cells compromise tissue homeostasis and function, and their presence has been causally implicated in the development of such age-associated conditions as sarcopenia and cataracts [92].

Several studies have demonstrated a role for innate immune cells in the recognition and clearance of senescent cells [93, 94]. Of interest, via granule exocytosis, NK cells elicit potent cytotoxicity towards senescent cells [11]. As NK cells from older adults exhibit impaired perforin release upon target cell stimulation [33], NK cell immunosenescence may be one mechanism by which to explain the accumulation of senescent cells in aged tissue [11].

An emerging field of research in the area of senescence is the development of senolytics, a class of drugs aimed at selectively eliminating senescent cells. Recently, it was reported that a single-dose of senolytic drugs significantly reduced the number of senescent cells in multiple tissues of aged mice, resulting in amongst other things, improved functionality and the delayed onset of age-related symptoms and pathologies [95]. Whilst holding promise for combating the effects of ageing, senolytics will not reverse age-related changes in the immune system. Thus, should future studies prove that NK cells from older adults exhibit reduced cytotoxicity towards senescent cells, then an alternative to senolytics could be the therapeutic enhancement of NKCC, an approach that would restore the physiological route of senescent cell removal.

1.3.2 Resolution of Inflammation

In a series of recent in vitro co-culture experiments, NK cells were shown to induce neutrophil apoptosis in a contact-dependent manner that required recognition of an as yet unidentified ligand on the neutrophil surface by the NK cell activatory receptor NKp46 [12]. This observation, combined with the ability of NK cells to lyse DCs and T cells via perforin-mediated cytotoxicity [13], suggests a role for NK cells in the resolution of inflammation.

Human ageing is associated with a reduced frequency of circulating NKp46$^+$ NK cells [33, 41]. Couple this change in surface phenotype to the reduced number of NK cells present at sites of inflammation in aged mice [36, 37], then it is conceivable to think that the rates of NK cell-mediated neutrophil apoptosis would be reduced in older adults, a consequence of which would be the slower resolution of inflammatory responses. Along similar lines, based on our data showing an age-related reduction in perforin secretion by activated NK cells [33], one would predict that NKCC towards CD4$^+$ T cells, which in the setting of viral infection has been proposed to prevent immune-mediated pathology [13], would also be reduced in older adults.

Away from NK cells, we recently provided *ex vivo* evidence that suggests ageing is associated with increased rates of neutrophil degranulation in the absence of infection [15]. Coinciding with elevated systemic neutrophil protease activity, this age–related increase in the function of unstimulated neutrophils may contribute to the chronic low-grade inflammatory state of older adults by triggering widespread collateral tissue damage [15].

1.3.3 Wound Healing

Critical for efficient wound healing is an appropriate inflammatory response, which is characterised in part by infiltration into the wound of neutrophils and macrophages. Upon arrival, neutrophils are primarily responsible for tissue debridement and the clearance of invading microbes, whilst macrophages promote cell recruitment and the removal of apoptotic neutrophils. Given the changes that occur in neutrophil and macrophage function with age, one would predict differences in the inflammatory phase of wound repair between young and older adults. Indeed, in a murine model of dermal injury, Swift *et al.* reported reduced phagocytic capacity for macrophages isolated from the wounds of aged mice [96], whilst in a cutaneous wound infection model, reduced neutrophil and macrophage numbers in the wounds of aged mice were detected despite no age-related difference in chemokine levels [19]. As neutrophils clear cellular debris, and macrophages, through the clearance of apoptotic neutrophils, reduce bystander tissue damage and initiate the resolution phase of inflammation, innate immunosenescence may be one explanation for the delay in wound healing that accompanies physiological ageing [14].

1.3.4   Inflammageing

Alongside the aforementioned changes in the number and function of innate immune cells, ageing is associated with a low-grade systemic up-regulation of circulating inflammatory mediators. Termed *inflammageing*, this sub-clinical chronic inflammatory state is characterised by elevated levels of pro-inflammatory cytokines (e.g. IL-6 and TNF-α) and acute phase proteins (e.g. C reactive protein (CRP)), as well as reduced levels of the anti-inflammatory cytokine IL-10 [97–100]. Whilst age-related changes in immune function undoubtedly contribute to the development of *inflammageing*, this phenomenon is driven by a multitude of age-associated factors, which include the accumulation of pro-inflammatory cytokine secreting senescent cells [90, 101], an increase in fat tissue [98] and the decline in sex steroid synthesis [102].

Results from longitudinal-based studies suggest *inflammageing* is deleterious to human health with studies in older cohorts demonstrating that low-grade increases in the circulating levels of TNF-α [103], IL-6 [103, 104] and CRP [105] are associated with both all-cause [103, 104] and cause-specific [104, 105] mortality. Furthermore, *inflammageing* is a predictor of frailty [106] and is considered a major factor in the development of several age-related pathologies, such as atherosclerosis [107], Alzheimer's disease [100] and sarcopenia [108].

1.4   Conclusions

Over the past decade, our understanding of innate immunosenescence and its underlying mechanisms has grown considerably, with altered cell signalling emerging as a prominent cause of many age-related changes in immune function [15, 17, 30, 31, 43]. Therapeutic manipulation of intracellular signalling pathways may therefore represent a means by which to counteract immunosenescence and enhance immune function in older adults. On this note, we recently proposed that targeting PI3K signalling may improve the migratory accuracy of neutrophils during infectious episodes in older adults [15], whilst inhibition of the sterol pathway has recently been shown to enhance NET formation, a neutrophil function that is significantly reduced with age [27, 28, 109]. As a field of research in its infancy, novel therapeutic targets for the "treatment" of immunosenescence will almost certainly emerge from future studies, with one potential candidate being myeloid derived suppressor cells (MDSCs). A heterogeneous population of cells with potent immune suppressive properties, the frequency of circulating MDSCs increases with age [110], which may contribute to the increased incidence of infection and cancer amongst older adults as well as the reduced efficacy of vaccinations in this group [111]. In other fields of research, the inhibition or elimination of MDSCs is currently being investigated as an approach by which to boost immune responses [112]. Results of such studies should be of interest to immune gerontologists as it may offer an insight into whether similar approaches may benefit older adults in terms of alleviating age-related immune dysfunction.

References

1. United Nations Department of Economic and Social Affairs Population Division. World Population Ageing 1950–2050. Executive Summary. http://www.un.org/esa/population/publications/worldageing19502050/ (2016). Accessed on 03 Jan 2016
2. Comfort A. Ageing: the biology of senescence. London: Routledge and Kegan Paul; 1964. p. 365.
3. Emori TG, Banerjee SN, Culver DH, Gaynes RP, Horan TC, Edwards JR, et al. Nosocomial infections in elderly patients in the United States, 1986–1990. National Nosocomial Infections Surveillance System. Am J Med. 1991;91:289S–93.
4. Fleming DM, Elliot AJ. The impact of influenza on the health and health care utilisation of elderly people. Vaccine. 2005;23 Suppl 1:S1–9.
5. Marston BJ, Plouffe JF, File Jr TM, Hackman BA, Salstrom SJ, Lipman HB, et al. Incidence of community-acquired pneumonia requiring hospitalization. Results of a population-based active surveillance Study in Ohio. The Community-Based Pneumonia Incidence Study Group. Arch Intern Med. 1997;157:1709–18.
6. Thompson WW, Shay DK, Weintraub E, Brammer L, Cox N, Anderson LJ, et al. Mortality associated with influenza and respiratory syncytial virus in the United States. JAMA. 2003;289:179–86.
7. Fein AM. Pneumonia in the elderly: overview of diagnostic and therapeutic approaches. Clin Infect Dis. 1999;28:726–9.
8. Aw D, Silva AB, Palmer DB. Immunosenescence: emerging challenges for an ageing population. Immunology. 2007;120:435–46.
9. Ho JC, Chan KN, Hu WH, Lam WK, Zheng L, Tiope GL, et al. The effect of aging on nasal mucociliary clearance, beat frequency, and ultrastructure of respiratory cilia. Am J Respir Crit Care Med. 2001;163:983–8.
10. Nagaki K, Hiramatsu S, Inai S, Sasaki A. The effect of aging on complement activity (CH50) and complement protein levels. J Clin Lab Immunol. 1980;3:45–50.
11. Sagiv A, Biran A, Yon M, Simon J, Lowe SW, Krizhanovsky V. Granule exocytosis mediates immune surveillance of senescent cells. Oncogene. 2013;32:1971–7.
12. Thoren FB, Riise RE, Ousback J, Della Chiesa M, Alsterholm M, Marcenaro E, et al. Human NK Cells induce neutrophil apoptosis via an NKp46- and Fas-dependent mechanism. J Immunol. 2012;188:1668–74.
13. Waggoner SN, Cornberg M, Selin LK, Welsh RM. Natural killer cells act as rheostats modulating antiviral T cells. Nature. 2012;481:394–8.
14. Guo S, DiPietro LA. Factors affecting wound healing. J Dent Res. 2010;89:219–29.
15. Sapey E, Greenwood H, Walton G, Mann E, Love A, Aaronson N, et al. Phosphoinositide 3-kinase inhibition restores neutrophil accuracy in the elderly: toward targeted treatments for immunosenescence. Blood. 2014;123:239–48.

16. Niwa Y, Kasama T, Miyachi Y, Kanoh T. Neutrophil chemotaxis, phagocytosis and parameters of reactive oxygen species in human aging: cross-sectional and longitudinal studies. Life Sci. 1989;44:1655–64.
17. Fulop T, Larbi A, Douziech N, Fortin C, Guerard KP, Lesur O, et al. Signal transduction and functional changes in neutrophils with aging. Aging Cell. 2004;3:217–26.
18. Biasi D, Bambara LM, Carletto A, Caramaschi P, Andrioli G, Urbani G, et al. Neutrophil migration, oxidative metabolism and adhesion in early onset periodontitis. J Clin Periodontol. 1999;26:563–8.
19. Brubaker AL, Rendon JL, Ramirez L, Choudhry MA, Kovacs EJ. Reduced neutrophil chemotaxis and infiltration contributes to delayed resolution of cutaneous wound infection with advanced age. J Immunol. 2013;190:1746–57.
20. Nacionales DC, Szpila B, Ungaro R, Lopez MC, Zhang J, Gentile LF, et al. A detailed characterization of the dysfunctional immunity and abnormal myelopoiesis induced by severe shock and trauma in the aged. J Immunol. 2015;195:2396–407.
21. Emanuelli G, Lanzio M, Anfossi T, Romano S, Anfossi G, Calcamuggi G. Influence of age on polymorphonuclear leukocytes in vitro: phagocytic activity in healthy human subjects. Gerontology. 1986;32:308–16.
22. Butcher SK, Chahal H, Nayak L, Sinclair A, Henriquez NV, Sapey E, et al. Senescence in innate immune responses: reduced neutrophil phagocytic capacity and CD16 expression in elderly humans. J Leukoc Biol. 2001;70:881–6.
23. Wenisch C, Patruta S, Daxbock F, Krause R, Horl W. Effect of age on human neutrophil function. J Leukoc Biol. 2000;67:40–5.
24. Simell B, Vuorela A, Ekstrom N, Palmu A, Reunanen A, Meri S, et al. Aging reduces the functionality of anti-pneumococcal antibodies and the killing of Streptococcus pneumoniae by neutrophil phagocytosis. Vaccine. 2011;29:1929–34.
25. Braga PC, Sala MT, Dal SM, Pecile A, Annoni G, Vergani C. Age-associated differences in neutrophil oxidative burst (chemiluminescence). Exp Gerontol. 1998;33:477–84.
26. Brinkmann V, Reichard U, Goosmann C, Fauler B, Uhlemann Y, Weiss DS, et al. Neutrophil extracellular traps kill bacteria. Science. 2004;303:1532–5.
27. Tseng CW, Kyme PA, Arruda A, Ramanujan VK, Tawackoli W, Liu GY. Innate immune dysfunctions in aged mice facilitate the systemic dissemination of methicillin-resistant S. aureus. PLoS One. 2012;7:e41454.
28. Hazeldine J, Harris P, Chapple IL, Grant M, Greenwood H, Livesey A, et al. Impaired neutrophil extracellular trap formation: a novel defect in the innate immune system of aged individuals. Aging Cell. 2014;13:690–8.
29. Bianchi M, Hakkim A, Brinkmann V, Siler U, Seger RA, Zychlinsky A, et al. Restoration of NET formation by gene therapy in CGD controls aspergillosis. Blood. 2009;114:2619–22.
30. Alvarez E, Ruiz-Gutierrez V, Sobrino F, Santa-Maria C. Age-related changes in membrane lipid composition, fluidity and respiratory burst in rat peritoneal neutrophils. Clin Exp Immunol. 2001;124:95–102.
31. Fortin CF, Lesur O, Fulop Jr T. Effects of aging on triggering receptor expressed on myeloid cells (TREM)-1-induced PMN functions. FEBS Lett. 2007;581:1173–8.
32. Zhang Y, Wallace DL, de Lara CM, Ghattas H, Asquith B, Worth A, et al. In vivo kinetics of human natural killer cells: the effects of ageing and acute and chronic viral infection. Immunology. 2007;121:258–65.
33. Hazeldine J, Hampson P, Lord JM. Reduced release and binding of perforin at the immunological synapse underlies the age-related decline in natural killer cell cytotoxicity. Aging Cell. 2012;11:751–9.
34. Lutz CT, Karapetyan A, Al-Attar A, Shelton BJ, Holt KJ, Tucker JH, et al. Human NK cells proliferate and die in vivo more rapidly than T cells in healthy young and elderly adults. J Immunol. 2011;186:4590–8.
35. Hayhoe RP, Henson SM, Akbar AN, Palmer DB. Variation of human natural killer cell phenotypes with age: identification of a unique KLRG1-negative subset. Hum Immunol. 2010;71:676–81.

36. Beli E, Clinthorne JF, Duriancik DM, Hwang I, Kim S, Gardner EM. Natural killer cell function is altered during the primary response of aged mice to influenza infection. Mech Ageing Dev. 2011;132:503–10.

37. Fang M, Roscoe F, Sigal LJ. Age-dependent susceptibility to a viral disease due to decreased natural killer cell numbers and trafficking. J Exp Med. 2010;207:2369–81.

38. Mariani E, Meneghetti A, Neri S, Ravaglia G, Forti P, Cattini L, et al. Chemokine production by natural killer cells from nonagenarians. Eur J Immunol. 2002;32:1524–9.

39. Smyth MJ, Cretney E, Kelly JM, Westwood JA, Street SE, Yagita H, et al. Activation of NK cell cytotoxicity. Mol Immunol. 2005;42:501–10.

40. Hazeldine J, Lord JM. The impact of ageing on natural killer cell function and potential consequences for health in older adults. Ageing Res Rev. 2013;12:1069–78.

41. Almeida-Oliveira A, Smith-Carvalho M, Porto LC, Cardoso-Oliveira J, Ribeiro Ados S, Falcao RR, et al. Age-related changes in natural killer cell receptors from childhood through old age. Hum Immunol. 2011;72:319–29.

42. Facchini A, Mariani E, Mariani AR, Papa S, Vitale M, Manzoli FA. Increased number of circulating Leu 11+ (CD 16) large granular lymphocytes and decreased NK activity during human ageing. Clin Exp Immunol. 1987;68:340–7.

43. Mariani E, Mariani AR, Meneghetti A, Tarozzi A, Cocco L, Facchini A. Age-dependent decreases of NK cell phosphoinositide turnover during spontaneous but not Fc-mediated cytolytic activity. Int Immunol. 1998;10:981–9.

44. Vitale M, Della CM, Carlomagno S, Pende D, Arico M, Moretta L, et al. NK-dependent DC maturation is mediated by TNFalpha and IFNgamma released upon engagement of the NKp30 triggering receptor. Blood. 2005;106:566–71.

45. Krishnaraj R, Bhooma T. Cytokine sensitivity of human NK cells during immunosenescence. 2. IL2-induced interferon gamma secretion. Immunol Let. 1996;50:59.

46. Mariani E, Pulsatelli L, Meneghetti A, Dolzani P, Mazzetti I, Neri S, et al. Different IL-8 production by T and NK lymphocytes in elderly subjects. Mech Ageing Dev. 2001;122:1383–95.

47. Chiu BC, Stolberg VR, Chensue SW. Mononuclear phagocyte-derived IL-10 suppresses the innate IL-12/IFN-gamma axis in lung-challenged aged mice. J Immunol. 2008;181:3156–66.

48. Biron CA, Byron KS, Sullivan JL. Severe herpesvirus infections in an adolescent without natural killer cells. N Engl J Med. 1989;320:1731–5.

49. Ogata K, Yokose N, Tamura H, An E, Nakamura K, Dan K, et al. Natural killer cells in the late decades of human life. Clin Immunol Immunopathol. 1997;84:269–75.

50. Ogata K, An E, Shioi Y, Nakamura K, Luo S, Yokose N, et al. Association between natural killer cell activity and infection in immunologically normal elderly people. Clin Exp Immunol. 2001;124:392–7.

51. Levy SM, Herberman RB, Lee J, Whiteside T, Beadle M, Heiden L, et al. Persistently low natural killer cell activity, age, and environmental stress as predictors of infectious morbidity. Nat Immun Cell Growth Regul. 1991;10:289–307.

52. Mysliwska J, Trzonkowski P, Szmit E, Brydak LB, Machala M, Mysliwski A. Immunomodulating effect of influenza vaccination in the elderly differing in health status. Exp Gerontol. 2004;39:1447–58.

53. Wong KL, Yeap WH, Tai JJ, Ong SM, Dang TM, Wong SC. The three human monocyte subsets: implications for health and disease. Immunol Res. 2012;53:41–57.

54. Qian F, Wang X, Zhang L, Chen S, Piecychna M, Allore H, et al. Age-associated elevation in TLR5 leads to increased inflammatory responses in the elderly. Aging Cell. 2012;11:104–10.

55. Seidler S, Zimmermann HW, Bartneck M, Trautwein C, Tacke F. Age-dependent alterations of monocyte subsets and monocyte-related chemokine pathways in healthy adults. BMC Immunol. 2010;11:30.

56. Hearps AC, Martin GE, Angelovich TA, Cheng WJ, Maisa A, Landay AL, et al. Aging is associated with chronic innate immune activation and dysregulation of monocyte phenotype and function. Aging Cell. 2012;11:867–75.
57. Nyugen J, Agrawal S, Gollapudi S, Gupta S. Impaired functions of peripheral blood monocyte subpopulations in aged humans. J Clin Immunol. 2010;30:806–13.
58. Takeda K, Kaisho T, Akira S. Toll-like receptors. Annu Rev Immunol. 2003;21:335–76.
59. van Duin D, Mohanty S, Thomas V, Ginter S, Montgomery RR, Fikrig E, et al. Age-associated defect in human TLR-1/2 function. J Immunol. 2007;178:970–5.
60. Lim JS, Nguyen KC, Nguyen CT, Jang IS, Han JM, Fabian C, et al. Flagellin-dependent TLR5/caveolin-1 as a promising immune activator in immunosenescence. Aging Cell. 2015;14:907–15.
61. Taylor DN, Treanor JJ, Strout C, Johnson C, Fitzgerald T, Kavita U, et al. Induction of a potent immune response in the elderly using the TLR-5 agonist, flagellin, with a recombinant hemagglutinin influenza-flagellin fusion vaccine (VAX125, STF2.HA1 SI). Vaccine. 2011;29:4897–902.
62. van Duin D, Allore HG, Mohanty S, Ginter S, Newman FK, Belshe RB, et al. Prevaccine determination of the expression of costimulatory B7 molecules in activated monocytes predicts influenza vaccine responses in young and older adults. J Infect Dis. 2007;195:1590–7.
63. Olivieri F, Procopio AD, Montgomery RR. Effect of aging on microRNAs and regulation of pathogen recognition receptors. Curr Opin Immunol. 2014;29:29–37.
64. Renshaw M, Rockwell J, Engleman C, Gewirtz A, Katz J, Sambhara S. Cutting edge: impaired Toll-like receptor expression and function in aging. J Immunol. 2002; 169:4697–701.
65. Boehmer ED, Goral J, Faunce DE, Kovacs EJ. Age-dependent decrease in Toll-like receptor 4-mediated proinflammatory cytokine production and mitogen-activated protein kinase expression. J Leukoc Biol. 2004;75:342–9.
66. Boehmer ED, Meehan MJ, Cutro BT, Kovacs EJ. Aging negatively skews macrophage. Mech Ageing Dev. 2005;126:1305–13.
67. Chelvarajan RL, Collins SM, Van Willigen JM, Bondada S. The unresponsiveness of aged mice to polysaccharide antigens is a result of a defect in macrophage function. J Leukoc Biol. 2005;77:503–12.
68. Corsini E, Battaini F, Lucchi L, Marinovich M, Racchi M, Govoni S, et al. A defective protein kinase C anchoring system underlying age-associated impairment in TNF-alpha production in rat macrophages. J Immunol. 1999;163:3468–73.
69. Jiang M, Xiang Y, Wang D, Gao J, Liu D, Liu Y, et al. Dysregulated expression of miR-146a contributes to age-related dysfunction of macrophages. Aging Cell. 2012;11:29–40.
70. Agius E, Lacy KE, Vukmanovic-Stejic M, Jagger AL, Papageorgiou AP, Hall S, et al. Decreased TNF-alpha synthesis by macrophages restricts cutaneous immunosurveillance by memory CD4+ T cells during aging. J Exp Med. 2009;206:1929–40.
71. Agrawal A, Agrawal S, Cao JN, Su H, Osann K, Gupta S. Altered innate immune functioning of dendritic cells in elderly humans: a role of phosphoinositide 3-kinase-signaling pathway. J Immunol. 2007;178:6912–22.
72. Jing Y, Shaheen E, Drake RR, Chen N, Gravenstein S, Deng Y. Aging is associated with a numerical and functional decline in plasmacytoid dendritic cells, whereas myeloid dendritic cells are relatively unaltered in human peripheral blood. Hum Immunol. 2009;70:777–84.
73. Panda A, Qian F, Mohanty S, Van Duin D, Newman FK, Zhang L, et al. Age-associated decrease in TLR function in primary human dendritic cells predicts influenza vaccine response. J Immunol. 2010;184:2518–27.
74. Canaday DH, Amponsah NA, Jones L, Tisch DJ, Hornick TR, Ramachandra L. Influenza-induced production of interferon-alpha is defective in geriatric individuals. J Clin Immunol. 2010;30:373–83.
75. Della BS, Bierti L, Presicce P, Arienti R, Valenti M, Saresella M, et al. Peripheral blood dendritic cells and monocytes are differently regulated in the elderly. Clin Immunol. 2007;122:220–8.

76. Bhushan M, Cumberbatch M, Dearman RJ, Andrew SM, Kimber I, Griffiths CE. Tumour necrosis factor-alpha-induced migration of human Langerhans cells: the influence of ageing. Br J Dermatol. 2002;146:32–40.

77. Shodell M, Siegal FP. Circulating, interferon-producing plasmacytoid dendritic cells decline during human ageing. Scand J Immunol. 2002;56:518–21.

78. Agrawal A, Gupta S. Impact of aging on dendritic cell functions in humans. Ageing Res Rev. 2011;10:336–45.

79. Sridharan A, Esposo M, Kaushal K, Tay J, Osann K, Agrawal S, et al. Age-associated impaired plasmacytoid dendritic cell functions lead to decreased CD4 and CD8 T cell immunity. Age (Dordr). 2011;33:363–76.

80. Stout-Delgado HW, Yang X, Walker WE, Tesar BM, Goldstein DR. Aging impairs IFN regulatory factor 7 up-regulation in plasmacytoid dendritic cells during TLR9 activation. J Immunol. 2008;181:6747–56.

81. Grolleau-Julius A, Harning EK, Abernathy LM, Yung RL. Impaired dendritic cell function in aging leads to defective antitumor immunity. Cancer Res. 2008;68:6341–9.

82. Zhao J, Zhao J, Legge K, Perlman S. Age-related increases in PGD(2) expression impair respiratory DC migration, resulting in diminished T cell responses upon respiratory virus infection in mice. J Clin Invest. 2011;121:4921–30.

83. Liu WM, Nahar TE, Jacobi RH, Gijzen K, Van Beek J, Hak E, et al. Impaired production of TNF-alpha by dendritic cells of older adults leads to a lower CD8+ T cell response against influenza. Vaccine. 2012;30:1659–66.

84. Donnini A, Argentati K, Mancini R, Smorlesi A, Bartozzi B, Bernardini G, et al. Phenotype, antigen-presenting capacity, and migration of antigen-presenting cells in young and old age. Exp Gerontol. 2002;37:1097–112.

85. Tesar BM, Walker WE, Unternaehrer J, Joshi NS, Chandele A, Haynes L, et al. Murine [corrected] myeloid dendritic cell-dependent toll-like receptor immunity is preserved with aging. Aging Cell. 2006;5:473–86.

86. Grolleau-Julius A, Garg MR, Mo R, Stoolman LL, Yung RL. Effect of aging on bone marrow-derived murine CD11c+CD4-CD8alpha- dendritic cell function. J Gerontol A Biol Sci Med Sci. 2006;61:1039–47.

87. Wong CP, Magnusson KR, Ho E. Aging is associated with altered dendritic cells subset distribution and impaired proinflammatory cytokine production. Exp Gerontol. 2010;45:163–9.

88. Ferlazzo G, Tsang ML, Moretta L, Melioli G, Steinman RM, Munz C. Human dendritic cells activate resting natural killer (NK) cells and are recognized via the NKp30 receptor by activated NK cells. J Exp Med. 2002;195:343–51.

89. Brancato SK, Albina JE. Wound macrophages as key regulators of repair: origin, phenotype, and function. Am J Pathol. 2011;178:19–25.

90. Dimri GP, Lee X, Basile G, Acosta M, Scott G, Roskelley C, et al. A biomarker that identifies senescent human cells in culture and in aging skin in vivo. Proc Natl Acad Sci U S A. 1995;92:9363–7.

91. Price JS, Waters JG, Darrah C, Pennington C, Edwards DR, Donell ST, et al. The role of chondrocyte senescence in osteoarthritis. Aging Cell. 2002;1:57–65.

92. Baker DJ, Wijshake T, Tchkonia T, LeBrasseur NK, Childs BG, van de Sluis B, et al. Clearance of p16Ink4a-positive senescent cells delays ageing-associated disorders. Nature. 2011;479:232–6.

93. Krizhanovsky V, Yon M, Dickins RA, Hearn S, Simon J, Miething C, et al. Senescence of activated stellate cells limits liver fibrosis. Cell. 2008;134:657–67.

94. Xue W, Zender L, Miething C, Dickins RA, Hernando E, Krizhanovsky V, et al. Senescence and tumour clearance is triggered by p53 restoration in murine liver carcinomas. Nature. 2007;445:656–60.

95. Zhu Y, Tchkonia T, Pirtskhalava T, Gower AC, Ding H, Giorgadze N, et al. The Achilles' heel of senescent cells: from transcriptome to senolytic drugs. Aging Cell. 2015;14:644–58.

96. Swift ME, Burns AL, Gray KL, DiPietro LA. Age-related alterations in the inflammatory response to dermal injury. J Invest Dermatol. 2001;117:1027–35.

97. Hager K, Machein U, Krieger S, Platt D, Seefried G, Bauer J. Interleukin-6 and selected plasma proteins in healthy persons of different ages. Neurobiol Aging. 1994;15:771–2.
98. Pedersen M, Bruunsgaard H, Weis N, Hendel HW, Andreassen BU, Eldrup E, et al. Circulating levels of TNF-alpha and IL-6-relation to truncal fat mass and muscle mass in healthy elderly individuals and in patients with type-2 diabetes. Mech Ageing Dev. 2003;124:495–502.
99. Lio D, Scola L, Crivello A, Colonna-Romano G, Candore G, Bonafe M, et al. Gender-specific association between -1082 IL-10 promoter polymorphism and longevity. Genes Immun. 2002;3:30–3.
100. Bruunsgaard H, Andersen-Ranberg K, Jeune B, Pedersen AN, Skinhoj P, Pedersen BK. A high plasma concentration of TNF-alpha is associated with dementia in centenarians. J Gerontol A Biol Sci Med Sci. 1999;54:M357–64.
101. Minamino T, Miyauchi H, Yoshida T, Ishida Y, Yoshida H, Komuro I. Endothelial cell senescence in human atherosclerosis: role of telomere in endothelial dysfunction. Circulation. 2002;105:1541–4.
102. Roggia C, Gao Y, Cenci S, Weitzmann MN, Toraldo G, Isaia G, et al. Up-regulation of TNF-producing T cells in the bone marrow: a key mechanism by which estrogen deficiency induces bone loss in vivo. Proc Natl Acad Sci U S A. 2001;98:13960–5.
103. Bruunsgaard H, Ladelund S, Pedersen AN, Schroll M, Jorgensen T, Pedersen BK. Predicting death from tumour necrosis factor-alpha and interleukin-6 in 80-year-old people. Clin Exp Immunol. 2003;132:24–31.
104. Harris TB, Ferrucci L, Tracy RP, Corti MC, Wacholder S, Ettinger Jr WH, et al. Associations of elevated interleukin-6 and C-reactive protein levels with mortality in the elderly. Am J Med. 1999;106:506–12.
105. Gussekloo J, Schaap MC, Frolich M, Blauw GJ, Westendorp RG. C-reactive protein is a strong but nonspecific risk factor of fatal stroke in elderly persons. Arterioscler Thromb Vasc Biol. 2000;20:1047–51.
106. Baylis D, Bartlett DB, Syddall HE, Ntani G, Gale CR, Cooper C, et al. Immune-endocrine biomarkers as predictors of frailty and mortality: a 10-year longitudinal study in community-dwelling older people. Age (Dordr). 2013;35:963–71.
107. Bruunsgaard H, Skinhoj P, Pedersen AN, Schroll M, Pedersen BK. Ageing, tumour necrosis factor-alpha (TNF-alpha) and atherosclerosis. Clin Exp Immunol. 2000;121:255–60.
108. Visser M, Pahor M, Taaffe DR, Goodpaster BH, Simonsick EM, Newman AB, et al. Relationship of interleukin-6 and tumor necrosis factor-alpha with muscle mass and muscle strength in elderly men and women: the Health ABC Study. J Gerontol A Biol Sci Med Sci. 2002;57:M326–32.
109. Chow OA, von Kockritz-Blickwede M, Bright AT, Hensler ME, Zinkernagel AS, Cogen AL, et al. Statins enhance formation of phagocyte extracellular traps. Cell Host Microbe. 2010;8:445–54.
110. Verschoor CP, Johnstone J, Millar J, Dorrington MG, Habibagahi M, Lelic A, et al. Blood CD33(+)HLA-DR(-) myeloid-derived suppressor cells are increased with age and a history of cancer. J Leukoc Biol. 2013;93:633–7.
111. Bueno V, Sant'Anna OA, Lord JM. Ageing and myeloid-derived suppressor cells: possible involvement in immunosenescence and age-related disease. Age (Dordr). 2014;36:9729.
112. Wesolowski R, Markowitz J, Carson III WE. Myeloid derived suppressor cells - a new therapeutic target in the treatment of cancer. J Immunother Cancer. 2013;1:10.
113. McLachlan JA, Serkin CD, Morrey KM, Bakouche O. Antitumoral properties of aged human monocytes. J Immunol. 1995;154:832–43.
114. Ashcroft GS, Horan MA, Ferguson MW. Aging alters the inflammatory and endothelial cell adhesion molecule profiles during human cutaneous wound healing. Lab Invest. 1998;78:47–58.
115. Herrero C, Sebastian C, Marques L, Comalada M, Xaus J, Valledor AF, et al. Immunosenescence of macrophages: reduced MHC class II gene expression. Exp Gerontol. 2002;37:389–94.

116. Hilmer SN, Cogger VC, Le Couteur DG. Basal activity of Kupffer cells increases with old age. J Gerontol A Biol Sci Med Sci. 2007;62:973–8.
117. Videla LA, Tapia G, Fernandez V. Influence of aging on Kupffer cell respiratory activity in relation to particle phagocytosis and oxidative stress parameters in mouse liver. Redox Rep. 2001;6:155–9.
118. Ortega E, Garcia JJ, de la Fuente M. Modulation of adherence and chemotaxis of macrophages by norepinephrine. Influence of ageing. Mol Cell Biochem. 2000;203:113–7.
119. Smallwood HS, Lopez-Ferrer D, Squier TC. Aging enhances the production of reactive oxygen species and bactericidal activity in peritoneal macrophages by upregulating classical activation pathways. Biochemistry. 2011;50:9911–22.

Effects of Ageing on Adaptive Immune Responses

2

Sian M. Henson

Abstract

Persistent viral infections, reduced vaccination responses, increased autoimmunity, and a rise in inflammatory syndromes all typify immune ageing. As lifespan continues to extend, the demographic shift towards an older population will highlight the need to understand the mechanisms that drive age-related immune dysfunction, and to identify strategies to improve immune responsiveness in older people. These changes can be in part attributed to the accumulation of highly differentiated senescent T cells, characterised by their decreased proliferative capacity and the activation of senescence signaling pathways, together with alterations in the functional competence of regulatory cells, allowing inflammation to go unchecked. Moreover these defects that account for the decline in immune responsiveness also contribute to an increased prevalence in autoimmunity, through the reshaping of the peripheral T cell repertoire. This chapter discusses how the age-associated remodelling of the immune system leads to a lack of stability and subsequent decline in immune function.

Keywords

T cell • Lymphocyte differentiation • Senescence • Ageing • CD28 • TCR • mTOR • Regulatory • Immunosuppressive • Inflammation

2.1 Introduction

Immune function declines as we age resulting in an increased susceptibility to new infections and re-activation of latent pathogens to which we were once immune [1, 2]. Paradoxically, this dampened immune responsiveness observed during immune

S.M. Henson (✉)

Centre for Microvascular Research, William Harvey Research Institute, Barts & The London School of Medicine and Dentistry, Queen Mary, University of London, Charterhouse Square, London EC1M 6BQ, UK

e-mail: s.henson@qmul.ac.uk

© Springer International Publishing Switzerland 2017

V. Bueno et al. (eds.), *The Ageing Immune System and Health*,

DOI 10.1007/978-3-319-43365-3_2

ageing is also associated with a low-grade chronic inflammation, termed 'inflammageing' [3, 4]. Although inflammation is critical for dealing with infections and tissue damage, inflammageing appears to be physiologically deleterious and predictive of all-cause mortality in multiple aged cohorts [5]. Immune senescence results from defects in different leukocyte populations, however the dysfunction is most profound in T cells [6, 7]. The responses of T cells from aged individuals are typically slower and of a lower magnitude than those of young individuals, whether the response is measured by proliferation [8], telomerase activity [8] or the induction of signalling events [9]. The T cell pool contains a number of functionally distinct subsets: CD4$^+$ T cells, CD8$^+$ T cells, and regulatory T cells together with non-conventional T cells, while not all equally affected by age, the overall T cell number does decline dramatically as a result of thymic atrophy [10, 11]. This reduced thymic output leads to the homeostatic expansion of peripheral T cells to regenerate the T cell pool, which together with the turnover of T cells in response to repeated antigenic stimulation eventually lead to the accumulation of oligoclonally expanded, functionally impaired T cells [1, 12]. Recent evidence suggests that the rise in homeostatic expansion during ageing may also be responsible for the abundance of memory-phenotype T cells specific for viral antigens in adults never previously infected [13]. Whatever the cause of these age-related changes to the T cell compartment, they all contribute to the inability of the aged immune system to respond to new antigenic challenge and mount effective responses following vaccination [14].

This chapter will examine how T cell memory is affected during ageing, the contribution of changes to the T cells themselves, as well as the consequence of an altered regulatory balance will be discussed, and will show how the generation of highly differentiated end-stage T cells contributes to age-associated disease.

2.2   Phenotypic Differentiation of T Cells During Ageing

There are numerous ways in which human T cell differentiation can be characterised [15, 16], the most common being listed in Table 2.1. However the most striking characteristic of highly differentiated T cells is the loss of the co-stimulatory molecules CD27 and CD28 and the re-expression of CD45RA, with CD8$^+$ T cells losing CD28 first followed by CD27 with the inverse being true for CD4$^+$ T cells [15, 17, 18]. Initially, it was thought that the loss of CD28 was the predominant factor controlling reduced activity in these cells [19]. However, highly differentiated T cells show considerable redundancy in co-stimulatory receptor usage, and alternative receptors, such as OX40 and 4-1BB can promote T cell activation in CD28$^-$CD8$^+$ populations [8, 20]. Highly differentiated T cells also increase during ageing with similar phenotypic changes occurring in both CD4$^+$ and CD8$^+$ T cells. However the rate at which these changes happen varies within each subset, with age-related changes being more pronounced on CD8$^+$ T cells, possibly due them exhibiting a greater homeostatic stability than CD4$^+$ T cells [21].

Highly differentiated T cells remain functional and secrete high levels of cytokines such as interferon-γ (IFNγ) and tumour necrosis factor (TNFα), together with high

Table 2.1 Phenotypic and functional characteristics of differentiated human T cells

	Early differentiated		Mid differentiated		Late differentiated	
Differentiation	CD45RA	+++	CD45RA	+/−	CD45RA	−/+
	CD27	+++	CD27	+	CD27	−
	CD28	+++	CD28	+/−	CD28	−
	CCR7	+++	CCR7	+++	CCR7	−
	CD62L	+++	CD62L	+++	CD62L	−
	CD57	+	CD57	++	CD57	+++
	Cytotoxicity	+	Cytotoxicity	++	Cytotoxicity	+++
	Proliferation	++	Proliferation	++	Proliferation	+/−
Senescence	KLRG1	+	KLRG1	++	KLRG1	+++
	Bcl-2	+++	Bcl-2	++	Bcl-2	+
	Telomere	+++	Telomere	++	Telomere	+
	Telomerase	+++	Telomerase	++	Telomerase	−
	P-Akt	+++	P-Akt	++	P-Akt	−
	P-mTORC1	++	P-mTORC1	+++	P-mTORC1	−
	ROS	+	ROS	++	ROS	+++

levels of granzyme B and perforin expression, indicating that they have the potential to mediate high cytotoxic activity [16, 22–25]. Furthermore, highly differentiated CD27⁻CD28⁻ T cells remain polyfunctional secreting interleukin-2 (IL-2), IFNγ and TNFα, and expressing CD40 ligand, to the same extent as less differentiated memory T cell populations [22, 26, 27]. However, these cells have reduced capacity to replicate after activation [17, 27] and are susceptible to apoptosis ex vivo but it is possible that CD27⁻CD28⁻ T cells may persist in vivo in the presence of appropriate survival signals [28]. Taken together this suggests that highly differentiated T cell populations have characteristics of short-lived effector T cells, namely, potent effector function and susceptibility to apoptosis [29].

In addition, highly differentiated T cells become less reliant on specific antigen for stimulation and more prone to activation through innate receptors. Including killer immunoglobulin-like receptors (KIRs), killer cell lectin-like receptors (KLRs), and the immunoglobulin-like transcript receptors (ILT/CD85), more typically associated with NK cell function [30]. The acquisition of these receptors on highly differentiated T cells is more frequently seen on CD8⁺ than CD4⁺ T cells [31]. The majority of these NK receptors recognise MHC class I molecules [32], thereby circumventing the need for antigen recognition, which restricts clonal expansion and offers protection from undue repertoire skewing. Interestingly, NK receptors are not the only innate molecules to increase on end-stage T cells, recent evidence suggests that Toll-like receptor (TLRs) are also upregulated. TLRs appear to be more important for CD4⁺ T cell function, displaying higher levels of TLR2/4 expression than CD8⁺ T cells [33]. However the pattern recognition molecule retinoic acid inducible gene I (RIG-I)-like helicase was found to be more important for highly differentiated CD8⁺ T cell function [33]. This suggests that differential expression of TLRs and RIG-1 on CD4⁺ and CD8⁺ T cells may reflect distinct modes of antigen-independent T cell priming.

2.3   Change in T Cell Signalling During Ageing

T cell differentiation is a highly complex process controlled not only by co-stimulation but also by the strength and duration of T cell receptor (TCR) signalling [34]. Nearly all TCR signalling pathways have been found altered during ageing (Fig. 2.1), with several studies suggesting aberrancies in early TCR mediated signalling events, such as changes to protein tyrosine phosphorylation, calcium mobilization and the translocation of protein kinase C to the plasma membrane [35–37]. In addition, there is also a decline in proximal and intermediate signalling events leading to decreased transcription factor activity, notably NF-kB and NF-AT [38]. Furthermore recent findings have shown the Lck pathway to be an important factor controlling T cell signalling [39]. Lck activity is regulated by two phosphotyrosine residues, Tyr394 which stabilises an open conformation and promotes kinase activity, and Tyr505 which results in a closed conformation decreasing activity. The lack of phosphorylation in either site can result in partial activation of lck [40]. The balance among these three pools of differentially phosphorylated lck is thought to determine the general level of enzymatic activity of lck in T cells [40]. Age-related changes in both Tyr394 and Tyr505 have been reported for CD4$^+$ T cells, both showing increased phosphorylation [36], confounding the authors as phosphorylation at these sites have opposing effects. However recent data suggests that the local concentration of active lck molecules is more important than the

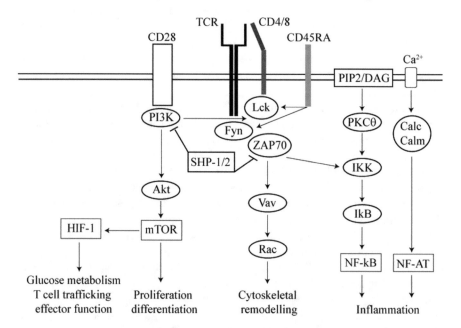

Fig. 2.1 T cell receptor signalling pathways. This scheme highlights the most investigated signalling pathways in the context of ageing research

overall phosphorylated state of lck [41]. Therefore it remains to be determined whether highly differentiated T cells exhibit changes in either the location or movement of lck molecules in the membrane.

The immune microenvironment also plays a crucial role in shaping lineage commitment and ultimately the function of T cells (Fig. 2.1). There is now much evidence that mTOR (mammalian target of rapamycin), plays a central role controlling T cell function through its ability to connect immune signalling to metabolism [42]. The kinase mTOR belongs to the phosphatidylinositol 3-kinase (PI3K)-related kinase (PIKK) family of proteins that act as regulators of cellular growth and metabolism [43]. mTOR is the catalytic subunit of two distinct signalling complexes, mTOR complex 1 (mTORC1) and mTOR complex 2 (mTORC2), the activity of which are differentially regulated by distinct accessory proteins [42, 43]. Increased mTOR activity has been linked to senescence and ageing [44–47], however there is growing evidence to suggest that this increase is not universally observed in every cell type nor is it evident in older humans [27, 48–50]. Highly differentiated CD4$^+$ (unpublished observations) and CD8$^+$ T cells are unable to phosphorylate either the mTOCR1 [27] or the mTORC2 complex [51].

mTORC1 activity is a requirement for the generation of effector molecules [52], and despite highly differentiated CD8$^+$ T cells apparent lack of mTOR activity they are highly potent effectors [26, 27]. mTORC1 has also been shown to control transcriptional programs that determine CD8$^+$ effector fate via a HIF1-dependent mechanism [53]. However HIF1-null CD8$^+$ T cells were shown to have many characteristics of effector CD8$^+$ T cells such as high levels of IFNγ production, they lacked perforin and granzyme expression. A situation which is mirrored in highly differentiated CD8$^+$ T cells isolated from old individuals, where these cells display high levels of TNFα and IFNγ but lower levels of perforin and granzyme [25]. Furthermore when the mTORC1 inhibitor rapamycin was incubated with highly differentiated CD8$^+$ T cells it had no effect on IFNγ production [54]. Thus corroborating with the idea that mTOR may not play a role in T cell senescence.

An increasingly recognized pathway for modulating T cell signalling is via reactive oxygen species (ROS; Fig. 2.1) [55]. ROS influences the balance between protein tyrosine kinase and phosphatase activities through redox-dependent regulation of signalling [38]. During TCR stimulation there is a transient increase in ROS, which inactivates SHP-1 facilitating TCR signalling, however in continued ROS SHP-1 regulation is further altered, leading to the negative regulation of TCR function. High levels of ROS have been found in highly differentiated CD8$^+$ T cells [27] and during ageing [56], generated in part by impaired mitochondrial function [27]. Indeed the requirement for robust mitochondria in antigen-specific T cell expansion has been demonstrated using mice with T cell-specific alterations to complex III [57]. Thus, ageing increased oxidative stress, together with changes in tyrosine kinase and phosphatase activities all contribute to the altered T cell signalling observed during T cell differentiation and ageing.

2.4   Altered Regulatory Capacity During Ageing

Immune function is also controlled by regulatory T and B cells and myeloid-derived suppressor cells (MDSCs), with emerging data suggesting that the functional competence of these regulatory cells is altered during ageing (Fig. 2.2). The most characterised of the regulatory subsets are the natural regulatory T cells (nTregs), defined as CD4$^+$CD25^{hi}Foxp3$^+$ [58]. They are fundamental for maintaining peripheral tolerance and protection against autoimmunity, and also modulating immunity to infections and tumors [58]. Therefore, to maintain controlled immunity, it is important that this regulatory population is maintained throughout life. CD4$^+$ nTregs are derived from the thymus, however, the thymus involutes with age [10] suggesting that nTreg numbers might also be reduced during aging. However, there is ample data to suggest that nTreg numbers increase with age [59–62]. Therefore the number of nTregs must either be sustained through extensive proliferation or by generation from an extra-thymic source [60, 63]. Human nTregs can express either CD45RA or CD45RO,

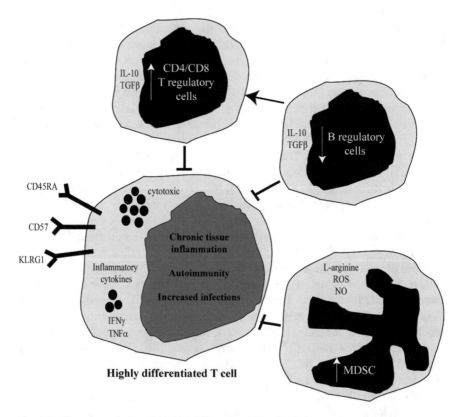

Fig. 2.2 Altered regulation of highly differentiated T cells during ageing. Tregs and MDSCs increase with age whereas Breg numbers have been shown to decline, all alter immune control leading to a loss of peripheral tolerance and increased autoimmunity, as well as modulating immune responses to infections

with 90–95 % of adult nTregs displaying a CD45RO$^+$ phenotype [64]. It has been shown that human CD45RO$^+$ Tregs represent a highly differentiated population, characterised by short telomeres, a loss of telomerase activity and an increased susceptibility to apoptosis [60]. Indicating that these cells have limited capacity for self-renewal, suggesting that the Treg pool is not maintained through continuous turn-over of pre-existing CD45RO$^+$ Tregs. Tregs can also be induced from CD4$^+$CD25$^-$Foxp3$^-$ conventional T cells in response to specific signals, such as TGFβ and retinoic acid, these so called inducible Tregs (iTregs) display the same suppressive function as nTregs [65]. It has been shown that human Tregs share a close TCR homology with CD4$^+$CD25$^-$ responder T cells [63, 66], supporting the hypothesis that the process of peripheral conversion from non-Tregs plays a significant role in the maintenance of the Treg population in humans. However mouse studies do not support this idea, when using non-immunised and non-lymphopenic mice, conversion was found not to play a significant role in the shaping of the peripheral Treg repertoire [67, 68]. Furthermore the induction of iTregs from aged mice was shown to be impaired [69]. However, the situation is likely to be different in humans, who undergo recurrent immunological challenges and have a much longer life span [65, 70].

Tregs belonging to the CD8$^+$ T cell compartment are equally important in regulating immune responses, although they are less well characterised than their CD4$^+$ counterparts [71]. Like CD4$^+$ Tregs, the percentage of CD8$^+$Foxp3$^+$ Tregs significantly increases in older individuals, with suppressor function remaining comparable to younger individuals [72]. Interestingly, these CD8$^+$ Tregs lacked expression of CD28, as discussed earlier the loss of CD28 is a hallmark of ageing. Thus suggesting that the increase in CD8$^+$Foxp3$^+$CD28$^-$ Tregs is consistent with the increase in overall numbers of CD8$^+$CD28$^-$ T cells.

Over the past decade, a population of immunosuppressive B cells or Bregs have come to prominence, having been shown to inhibit excessive inflammation [73]. Bregs function primarily by skewing T cell differentiation in favour of a regulatory phenotype in both mice [74] and humans [75], controlling Treg induction through direct cognate interactions between Bregs and T cells [76, 77]. Bregs can also suppress the expansion of pathogenic T cells through the production of IL-10, IL-35 and TGFβ [78]. Although the expression of IL-10 has been used to define populations of Bregs, many different surface markers have been used, leading to inherent problems in Breg subset definition, reviewed by Rosser and Mauri [73]. However two phenotypically distinct subsets of B cells: transitional CD19$^+$CD24^{hi}CD38^{hi} B cells [79] and CD19$^+$CD5$^+$CD1d^{hi} 'B10' B cells [80] have been demonstrated to exert immunosuppressive functions. The frequency and function of both these Breg subsets declines with age, owing to reduced CD4$^+$ T cell helper activity [81, 82]. The ability of Bregs isolated from old individuals to produce IL-10 following either ex vivo maturation or stimulation was also found to be reduced, and was linked to both impaired B cell signalling through CD40 and reduced expression of CD40L on CD4$^+$ T cells [81].

More recently myeloid-derived suppressor cells (MDSCs) have also been recognised as a population of immunosuppressive myeloid lineage cells capable of suppressing T cell functions [83]. MDSCs have been characterised in mice as belonging to either a monocytic, CD11b$^+$Ly6G$^-$Ly6C^{high} or granulocytic, CD11b$^+$LyG6$^+$Ly6C^{low}

lineage, with an analogous population being identified in humans, defined as CD33+HLA-DR− and lineage (CD3, CD19, CD56)-negative [84]. Reports have also shown human MDSCs to express CD11b and like their mouse counterparts have been subdivided into monocytic, CD14+ and granulocytic, CD15+ subtypes [85]. MDSCs suppress T cell responses principally through their ability to manipulate L-arginine metabolism, MDSCs produce arginase I, which catabolises L-arginine depriving T cells of this amino acid [83]. The loss of L-arginine from T cells in vitro inhibits their proliferation by arresting them in G_0/G_1 [86]. MDSCs can also inhibit CD8+ T cells through the production of ROS and peroxynitrite, which catalyze the nitration of the TCR, thereby preventing T cell-MHC interactions [87]. Furthermore MDSC can indirectly effect T cell activation through the induction of Tregs, this requires MDSC production of IL-10 and arginase, but depending on the subpopulation of MDSCs can either be TGFβ dependent or independent [88]. The frequency of MDSCs increases in numerous cancers [89], chronic viral infections [90] and ageing [91]. The accumulation of MDSCs with age is thought to be driven by inflammation, the pro-inflammatory cytokines IL-1β and IL-6 and PGE_2 all being shown to induce the differentiation of MDSCs [91].

2.5   Conclusion

This chapter describes the many ways in which T cell responses are detrimentally affected by ageing, highlighting the intrinsic defects that occur to memory T cells and the extrinsic effects of altering the balance of regulatory cell activity, both limiting the responsiveness of T cells. The importance of understanding this interplay is underscored by highly differentiated T cells being found in high numbers not only during ageing but in chronic viral infections [92, 93], malignancies [94] and autoimmunity [31, 95]. Highly differentiated T cells have also been regarded as a causative factor in acute transplant rejection [96]. Additionally the low grade inflammatory state observed during ageing also plays a causal role in atherosclerosis [97] and type II diabetes [98]. The immune impairments in patients with chronic hyperglycemia resemble those seen during ageing, namely poor control of infections and reduced vaccination responses [99]. The ageing immune system in its attempt to endure and overcome the acquired defects may thus contribute to the development of an unstable state that predisposes to disease.

References

1. Akbar AN, Fletcher JM. Memory T cell homeostasis and senescence during aging. Curr Opin Immunol. 2005;17:480–5.
2. Nikolich-Zugich J. Ageing and life-long maintenance of T-cell subsets in the face of latent persistent infections. Nat Rev Immunol. 2008;8:512–22.
3. Everett H, McFadden G. Viruses and apoptosis: meddling with mitochondria. Virology. 2001;288:1–7.
4. Franceschi C, Bonafe M, Valensin S, Olivieri F, De Luca M, Ottaviani E, et al. Inflammaging. An evolutionary perspective on immunosenescence. Ann N Y Acad Sci. 2000;908:244–54.

5. Bruunsgaard H, Andersen-Ranberg K, Hjelmborg J, Pedersen BK, Jeune B. Elevated levels of tumor necrosis factor alpha and mortality in centenarians. Am J Med. 2003;115:278–83.
6. Akbar AN, Beverley PC, Salmon M. Will telomere erosion lead to a loss of T-cell memory? Nat Rev Immunol. 2004;4:737–43.
7. Pawelec G. Immunity and ageing in man. Exp Gerontol. 2006;41:1239–42.
8. Plunkett FJ, Franzese O, Finney HM, Fletcher JM, Belaramani LL, Salmon M, et al. The loss of telomerase activity in highly differentiated CD8+CD28-CD27- T cells is associated with decreased Akt (Ser473) phosphorylation. J Immunol. 2007;178:7710–9.
9. Larbi A, Dupuis G, Khalil A, Douziech N, Fortin C, Fulop Jr T. Differential role of lipid rafts in the functions of CD4+ and CD8+ human T lymphocytes with aging. Cell Signal. 2006;18:1017–30.
10. Douek DC, McFarland RD, Keiser PH, Gage EA, Massey JM, Haynes BF, et al. Changes in thymic function with age and during the treatment of HIV infection. Nature. 1998;396:690–5.
11. Linton PJ, Dorshkind K. Age-related changes in lymphocyte development and function. Nat Immunol. 2004;5:133–9.
12. Messaoudi I, Lemaoult J, Guevara-Patino JA, Metzner BM, Nikolich-Zugich J. Age-related CD8 T cell clonal expansions constrict CD8 T cell repertoire and have the potential to impair immune defense. J Exp Med. 2004;200:1347–58.
13. Su Laura F, Kidd Brian A, Han A, Kotzin Jonathan J, Davis MM. Virus-specific CD4+ memory-phenotype T cells are abundant in unexposed adults. Immunity. 2013;38:373–83.
14. Goronzy JJ, Fulbright JW, Crowson CS, Poland GA, O'Fallon WM, Weyand CM. Value of immunological markers in predicting responsiveness to influenza vaccination in elderly individuals. J Virol. 2001;75:12182–7.
15. Appay V, van Lier RA, Sallusto F, Roederer M. Phenotype and function of human T lymphocyte subsets: consensus and issues. Cytometry A. 2008;73:975–83.
16. Sallusto F, Geginat J, Lanzavecchia A. Central memory and effector memory T cell subsets: function, generation, and maintenance. Annu Rev Immunol. 2004;22:745–63.
17. Fletcher JM, Vukmanovic-Stejic M, Dunne PJ, Birch KE, Cook JE, Jackson SE, et al. Cytomegalovirus-specific CD4+ T cells in healthy carriers are continuously driven to replicative exhaustion. J Immunol. 2005;175:8218–25.
18. Amyes E, Hatton C, Montamat-Sicotte D, Gudgeon N, Rickinson AB, McMichael AJ, et al. Characterization of the CD4+ T cell response to Epstein-Barr virus during primary and persistent infection. J Exp Med. 2003;198:903–11.
19. Effros RB, Dagarag M, Spaulding C, Man J. The role of CD8+ T-cell replicative senescence in human aging. Immunol Rev. 2005;205:147–57.
20. Waller EC, McKinney N, Hicks R, Carmichael AJ, Sissons JG, Wills MR. Differential costimulation through CD137 (4-1BB) restores proliferation of human virus-specific "effector memory" (CD28(-) CD45RA(HI)) CD8(+) T cells. Blood. 2007;110:4360–6.
21. Czesnikiewicz-Guzik M, Lee WW, Cui D, Hiruma Y, Lamar DL, Yang ZZ, et al. T cell subset-specific susceptibility to aging. Clin Immunol. 2008;127:107–18.
22. Brenchley JM, Karandikar NJ, Betts MR, Ambrozak DR, Hill BJ, Crotty LE, et al. Expression of CD57 defines replicative senescence and antigen-induced apoptotic death of CD8+ T cells. Blood. 2003;101:2711–20.
23. Khan N, Shariff N, Cobbold M, Bruton R, Ainsworth JA, Sinclair AJ, et al. Cytomegalovirus seropositivity drives the CD8 T cell repertoire toward greater clonality in healthy elderly individuals. J Immunol. 2002;169:1984–92.
24. van Lier RA, ten Berge IJ, Gamadia LE. Human CD8(+) T-cell differentiation in response to viruses. Nat Rev Immunol. 2003;3:931–9.
25. Henson SM, Macaulay R, Riddell NE, Nunn CJ, Akbar AN. Blockade of PD-1 or p38 MAP kinase signaling enhances senescent human CD8(+) T-cell proliferation by distinct pathways. Eur J Immunol. 2015;45:1441–51.
26. Koch S, Larbi A, Derhovanessian E, Ozcelik D, Naumova E, Pawelec G. Multiparameter flow cytometric analysis of CD4 and CD8 T cell subsets in young and old people. Immun Ageing. 2008;5:6.

27. Henson SM, Lanna A, Riddell NE, Franzese O, Macaulay R, Griffiths SJ, et al. p38 signaling inhibits mTORC1-independent autophagy in senescent human CD8+ T cells. J Clin Invest. 2014;124:4004–16.
28. Akbar AN, Salmon M, Savill J, Janossy G. A possible role for bcl-2 in regulating T-cell memory—a 'balancing act' between cell death and survival. Immunol Today. 1993;14:526–32.
29. Akbar AN, Henson SM. Are senescence and exhaustion intertwined or unrelated processes that compromise immunity? Nat Rev Immunol. 2011;11:289–95.
30. Abedin S, Michel JJ, Lemster B, Vallejo AN. Diversity of NKR expression in aging T cells and in T cells of the aged: the new frontier into the exploration of protective immunity in the elderly. Exp Gerontol. 2005;40:537–48.
31. Goronzy JJ, Li G, Yang Z, Weyand CM. The janus head of T cell aging - autoimmunity and immunodeficiency. Front Immunol. 2013;4:131.
32. Lopez-Botet M, Bellon T. Natural killer cell activation and inhibition by receptors for MHC class I. Curr Opin Immunol. 1999;11:301–7.
33. Hammond T, Lee S, Watson MW, Flexman JP, Cheng W, Fernandez S, et al. Toll-like receptor (TLR) expression on CD4+ and CD8+ T-cells in patients chronically infected with hepatitis C virus. Cell Immunol. 2010;264:150–5.
34. Williams MA, Bevan MJ. Effector and memory CTL differentiation. Annu Rev Immunol. 2007;25:171–92.
35. Pahlavani MA. T cell signaling: effect of age. Front Biosci. 1998;3:D1120–33.
36. Garcia GG, Miller RA. Age-related changes in lck-Vav signaling pathways in mouse CD4 T cells. Cell Immunol. 2009;259:100–4.
37. Larbi A, Douziech N, Dupuis G, Khalil A, Pelletier H, Guerard KP, et al. Age-associated alterations in the recruitment of signal-transduction proteins to lipid rafts in human T lympho-cytes. J Leukoc Biol. 2004;75:373–81.
38. Fulop T, Le Page A, Garneau H, Azimi N, Baehl S, Dupuis G, et al. Aging, immunosenescence and membrane rafts: the lipid connection. Longev Healthspan. 2012;1:6.
39. Acuto O, Di Bartolo V, Michel F. Tailoring T-cell receptor signals by proximal negative feed-back mechanisms. Nat Rev Immunol. 2008;8:699–712.
40. Brownlie RJ, Zamoyska R. T cell receptor signalling networks: branched, diversified and bounded. Nat Rev Immunol. 2013;13:257–69.
41. Nika K, Soldani C, Salek M, Paster W, Gray A, Etzensperger R, et al. Constitutively active Lck kinase in T cells drives antigen receptor signal transduction. Immunity. 2010;32:766–77.
42. Powell JD, Pollizzi KN, Heikamp EB, Horton MR. Regulation of immune responses by mTOR. Annu Rev Immunol. 2012;30:39–68.
43. Laplante M, Sabatini DM. mTOR signaling in growth control and disease. Cell. 2012;149:274–93.
44. Vigneron A, Vousden KH. p53, ROS and senescence in the control of aging. Aging. 2010;2:471–4.
45. Korotchkina LG, Leontieva OV, Bukreeva EI, Demidenko ZN, Gudkov AV, Blagosklonny MV. The choice between p53-induced senescence and quiescence is determined in part by the mTOR pathway. Aging. 2010;2:344–52.
46. Kennedy AL, Morton JP, Manoharan I, Nelson DM, Jamieson NB, Pawlikowski JS, et al. Activation of the PIK3CA/AKT pathway suppresses senescence induced by an activated RAS oncogene to promote tumorigenesis. Mol Cell. 2011;42:36–49.
47. Iglesias-Bartolome R, Patel V, Cotrim A, Leelahavanichkul K, Molinolo AA, Mitchell JB, et al. mTOR inhibition prevents epithelial stem cell senescence and protects from radiation-induced mucositis. Cell Stem Cell. 2012;11:401–14.
48. Arnold CR, Pritz T, Brunner S, Knabb C, Salvenmoser W, Holzwarth B, et al. T cell receptor-mediated activation is a potent inducer of macroautophagy in human CD8 + CD28+ T cells but not in CD8 + CD28− T cells. Exp Gerontol. 2014;54:75–83.
49. Harries LW, Fellows AD, Pilling LC, Hernandez D, Singleton A, Bandinelli S, et al. Advancing age is associated with gene expression changes resembling mTOR inhibition: evidence from two human populations. Mech Ageing Dev. 2012;133:556–62.

50. Fellows AD, Holly AC, Pilling LC, Melzer D, Harries LW. Age related changes in mTOR-related gene expression in two primary human cell lines. Healthy Aging Res. 2012;1:3.
51. Franzese O, Henson SM, Naro C, Bonmassar E. Defect in HSP90 expression in highly differentiated human CD8+ T lymphocytes. Cell Death Dis. 2014;5, e1294.
52. Rao RR, Li Q, Odunsi K, Shrikant PA. The mTOR kinase determines effector versus memory CD8+ T cell fate by regulating the expression of transcription factors T-bet and Eomesodermin. Immunity. 2010;32:67–78.
53. Finlay DK, Rosenzweig E, Sinclair LV, Feijoo-Carnero C, Hukelmann JL, Rolf J, et al. PDK1 regulation of mTOR and hypoxia-inducible factor 1 integrate metabolism and migration of CD8+ T cells. J Exp Med. 2012;209:2441–53.
54. Henson SM. CD8+ T-cell senescence: no role for mTOR. Biochem Soc Trans. 2015;43:734–9.
55. Torrão RC, Bennett SJ, Brown JE, Griffiths HR. Does metabolic reprogramming underpin age-associated changes in T cell phenotype and function? Free Radic Biol Med. 2014;71:26–35.
56. Pararasa C, Ikwuobe J, Shigdar S, Boukouvalas A, Nabney IT, Brown JE, et al. Age-associated changes in long-chain fatty acid profile during healthy aging promote pro-inflammatory monocyte polarization via PPARγ. Aging Cell. 2015:n/a-n/a.
57. Sena Laura A, Li S, Jairaman A, Prakriya M, Ezponda T, Hildeman David A, et al. Mitochondria are required for antigen-specific T cell activation through reactive oxygen species signaling. Immunity. 2013;38:225–36.
58. Sakaguchi S, Yamaguchi T, Nomura T, Ono M. Regulatory T cells and immune tolerance. Cell. 2008;133:775–87.
59. Lages CS, Suffia I, Velilla PA, Huang B, Warshaw G, Hildeman DA, et al. Functional regulatory T cells accumulate in aged hosts and promote chronic infectious disease reactivation. J Immunol. 2008;181:1835–48.
60. Vukmanovic-Stejic M, Zhang Y, Cook JE, Fletcher JM, McQuaid A, Masters JE, et al. Human CD4+ CD25hi Foxp3+ regulatory T cells are derived by rapid turnover of memory populations in vivo. J Clin Invest. 2006;116:2423–33.
61. Gregg R, Smith CM, Clark FJ, Dunnion D, Khan N, Chakraverty R, et al. The number of human peripheral blood CD4+ CD25high regulatory T cells increases with age. Clin Exp Immunol. 2005;140:540–6.
62. Raynor J, Lages CS, Shehata H, Hildeman DA, Chougnet CA. Homeostasis and function of regulatory T cells in aging. Curr Opin Immunol. 2012;24(4):482–7.
63. Booth NJ, McQuaid AJ, Sobande T, Kissane S, Agius E, Jackson SE, et al. Different proliferative potential and migratory characteristics of human CD4+ regulatory T cells that express either CD45RA or CD45RO. J Immunol. 2010;184:4317–26.
64. Taams LS, Smith J, Rustin MH, Salmon M, Poulter LW, Akbar AN. Human anergic/suppressive CD4(+)CD25(+) T cells: a highly differentiated and apoptosis-prone population. Eur J Immunol. 2001;31:1122–31.
65. Feuerer M, Hill JA, Mathis D, Benoist C. Foxp3+ regulatory T cells: differentiation, specification, subphenotypes. Nat Immunol. 2009;10:689–95.
66. Fazilleau N, Bachelez H, Gougeon ML, Viguier M. Cutting edge: size and diversity of CD4 + CD25high Foxp3+ regulatory T cell repertoire in humans: evidence for similarities and partial overlapping with CD4 + CD25- T cells. J Immunol. 2007;179:3412–6.
67. Wong J, Mathis D, Benoist C. TCR-based lineage tracing: no evidence for conversion of conventional into regulatory T cells in response to a natural self-antigen in pancreatic islets. J Exp Med. 2007;204:2039–45.
68. Pacholczyk R, Ignatowicz H, Kraj P, Ignatowicz L. Origin and T cell receptor diversity of Foxp3 + CD4 + CD25+ T cells. Immunity. 2006;25:249–59.
69. Chougnet CA, Tripathi P, Lages CS, Raynor J, Sholl A, Fink P, et al. A major role for Bim in regulatory T cell homeostasis. J Immunol. 2011;186:156–63.
70. Akbar AN, Vukmanovic-Stejic M, Taams LS, Macallan DC. The dynamic co-evolution of memory and regulatory CD4+ T cells in the periphery. Nat Rev Immunol. 2007;7:231–7.

71. Jagger A, Shimojima Y, Goronzy JJ, Weyand CM. Regulatory T cells and the immune aging process: a mini-review. Gerontology. 2014;60:130–7.
72. Simone R, Zicca A, Saverino D. The frequency of regulatory CD3+CD8+CD28- CD25+ T lymphocytes in human peripheral blood increases with age. J Leukoc Biol. 2008;84:1454–61.
73. Rosser EC, Mauri C. Regulatory B cells: origin, phenotype, and function. Immunity. 2015;42:607–12.
74. Carter NA, Vasconcellos R, Rosser EC, Tulone C, Munoz-Suano A, Kamanaka M, et al. Mice lacking endogenous IL-10-producing regulatory B cells develop exacerbated disease and present with an increased frequency of Th1/Th17 but a decrease in regulatory T cells. J Immunol. 2011;186:5569–79.
75. Flores-Borja F, Bosma A, Ng D, Reddy V, Ehrenstein MR, Isenberg DA, et al. CD19+CD24hiCD38hi B cells maintain regulatory T cells while limiting TH1 and TH17 differentiation. Sci Transl Med 2013;5:173ra23.
76. Rosser EC, Oleinika K, Tonon S, Doyle R, Bosma A, Carter NA, et al. Regulatory B cells are induced by gut microbiota-driven interleukin-1beta and interleukin-6 production. Nat Med. 2014;20:1334–9.
77. Yoshizaki A, Miyagaki T, DiLillo DJ, Matsushita T, Horikawa M, Kountikov EI, et al. Regulatory B cells control T-cell autoimmunity through IL-21-dependent cognate interactions. Nature. 2012;491:264–8.
78. Matsumoto M, Baba A, Yokota T, Nishikawa H, Ohkawa Y, Kayama H, et al. Interleukin-10-producing plasmablasts exert regulatory function in autoimmune inflammation. Immunity. 2014;41:1040–51.
79. Blair PA, Norena LY, Flores-Borja F, Rawlings DJ, Isenberg DA, Ehrenstein MR, et al. CD19(+) CD24(hi)CD38(hi) B cells exhibit regulatory capacity in healthy individuals but are functionally impaired in systemic Lupus Erythematosus patients. Immunity. 2010;32:129–40.
80. Yanaba K, Bouaziz JD, Haas KM, Poe JC, Fujimoto M, Tedder TF. A regulatory B cell subset with a unique CD1dhiCD5+ phenotype controls T cell-dependent inflammatory responses. Immunity. 2008;28:639–50.
81. Duggal NA, Upton J, Phillips AC, Sapey E, Lord JM. An age-related numerical and functional deficit in CD19+CD24hiCD38hiB cells is associated with an increase in systemic autoimmunity. Aging Cell. 2013;12:873–81.
82. Duggal NA, Upton J, Phillips AC, Lord JM. Development of depressive symptoms post hip fracture is associated with altered immunosuppressive phenotype in regulatory T and B lymphocytes. Biogerontology. 2015;17(1):229–39.
83. Gabrilovich DI, Nagaraj S. Myeloid-derived suppressor cells as regulators of the immune system. Nat Rev Immunol. 2009;9:162–74.
84. Lechner MG, Liebertz DJ, Epstein AL. Characterization of cytokine-induced myeloid-derived suppressor cells from normal human peripheral blood mononuclear cells. J Immunol. 2010;185:2273–84.
85. Greten TF, Manns MP, Korangy F. Myeloid derived suppressor cells in human diseases. Int Immunopharmacol. 2011;11:802–7.
86. Rodriguez PC, Quiceno DG, Ochoa AC. L-arginine availability regulates T-lymphocyte cell-cycle progression. Blood. 2007;109:1568–73.
87. Nagaraj S, Gupta K, Pisarev V, Kinarsky L, Sherman S, Kang L, et al. Altered recognition of antigen is a mechanism of CD8+ T cell tolerance in cancer. Nat Med. 2007;13:828–35.
88. Ostrand-Rosenberg S, Sinha P. Myeloid-derived suppressor cells: linking inflammation and cancer. J Immunol. 2009;182:4499–506.
89. Gabrilovich DI, Ostrand-Rosenberg S, Bronte V. Coordinated regulation of myeloid cells by tumours. Nat Rev Immunol. 2012;12:253–68.
90. Norris BA, Uebelhoer LS, Nakaya HI, Price AA, Grakoui A, Pulendran B. Chronic but not acute virus infection induces sustained expansion of myeloid suppressor cell numbers that inhibit viral-specific T cell immunity. Immunity. 2013;38:309–21.
91. Verschoor CP, Johnstone J, Millar J, Dorrington MG, Habibagahi M, Lelic A, et al. Blood CD33(+)HLA-DR(-) myeloid-derived suppressor cells are increased with age and a history of cancer. J Leukoc Biol. 2013;93:633–7.

92. Hertoghs KM, Moerland PD, van Stijn A, Remmerswaal EB, Yong SL, van de Berg PJ, et al. Molecular profiling of cytomegalovirus-induced human CD8+ T cell differentiation. J Clin Invest. 2010;120:4077–90.
93. Klenerman P, Hill A. T cells and viral persistence: lessons from diverse infections. Nat Immunol. 2005;6:873–9.
94. Perez-Mancera PA, Young ARJ, Narita M. Inside and out: the activities of senescence in cancer. Nat Rev Cancer. 2014;14:547–58.
95. Lindstrom TM, Robinson WH. Rheumatoid arthritis: a role for immunosenescence? J Am Geriatr Soc. 2010;58:1565–75.
96. Pawlik A, Florczak M, Masiuk M, Dutkiewicz G, Machalinski B, Rozanski J, et al. The expansion of CD4+CD28– t cells in patients with chronic kidney graft rejection. Transplant Proc. 2003; 35:2902–4.
97. Yu HT, Shin EC. T cell immunosenescence, hypertension, and arterial stiffness. Epidemiol Health. 2014;36, e2014005.
98. Pedicino D, Liuzzo G, Trotta F, Giglio AF, Giubilato S, Martini F, et al. Adaptive immunity, inflammation, and cardiovascular complications in type 1 and type 2 diabetes mellitus. J Diabetes Res. 2013;2013:184258.
99. Egawa Y, Ohfuji S, Fukushima W, Yamazaki Y, Morioka T, Emoto M, et al. Immunogenicity of influenza A(H1N1)pdm09 vaccine in patients with diabetes mellitus: with special reference to age, body mass index, and HbA1c. Hum Vaccin Immunother. 2014;10:1187–94.

The DNA Methylome: An Interface Between the Environment, Immunity, and Ageing

3

Lisa M. McEwen, Sarah J. Goodman, Michael S. Kobor, and Meaghan J. Jones

Abstract

The characteristic effects of ageing observed across the human lifespan are accompanied by a multitude of molecular changes. These age-related changes are a result of the complex interaction between our genetic makeup, lifestyle factors, and unique environments. People are subject to a variety of different exposures; many of these influences have the potential to "mark" our DNA and actually alter our cellular processes. This is a key component of epigenetics: a field that focuses on modifications to DNA and DNA packaging that function without altering the genetic sequence itself. DNA methylation is arguably the most well-characterized epigenetic modification, involving the addition of a methyl group to DNA, which, in an interesting paradox, is both stable long-term as well as plastic and reversible. DNA methylation fluctuates throughout the lifespan of mammalian organisms and has the potential to influence cellular processes through changes in gene expression. An important role of DNA methylation is as a molecular mediator between environmental exposures and physiological changes, which makes it a likely modifier of the immune system. In regards to the ageing process, the actual function of DNA methylation is unknown; however, global trends and site-specific changes in DNA methylation have been strongly correlated with chronological age. Here, we will discuss the particulars of epigenetics, with a focus on DNA methylation and its role in the development, maturation, dysfunction, and ageing of white blood cells of the immune system.

L.M. McEwen • S.J. Goodman • M.S. Kobor • M.J. Jones (✉)
Center for Molecular Medicine and Therapeutics, The University of British Columbia, BC Children's Hospital, A5-151 950 West 28th Ave, Vancouver, BC, Canada, V5Z 4H4
e-mail: mjones@cmmt.ubc.ca

© Springer International Publishing Switzerland 2017
V. Bueno et al. (eds.), *The Ageing Immune System and Health*,
DOI 10.1007/978-3-319-43365-3_3

Keywords
Epigenetics • DNA methylation • Epigenetic Drift • Epigenetic Clock • Age Acceleration • Biological Age • Blood • Immune system • Environment • Ageing

3.1   Introduction

Epigenetics is a rapidly emerging field investigating the interface between our genomes and our environments. Epigenetic mechanisms are responsible for the structural organization of DNA and its packaging proteins; one major function of which is to regulate gene expression by controlling DNA accessibility. While the genetic sequence is unaltered, these structural changes act as a regulatory stratum on the genome, instructing cellular identity and driving important cellular processes. Current characterized epigenetic marks include post-translational modifications to histone proteins, such as acetylation and methylation, histone variants, non-coding RNAs, and chemical modifications covalently bound directly to DNA. The latter include DNA methylation, the most commonly studied epigenetic mark, especially in human populations, which refers to the covalent attachment of a methyl group primarily to the 5′carbon of cytosine bases [1]. This chapter will briefly review DNA methylation, then examine the role of DNA methylation in three important areas: cell type specification, environmental embedding, and disease; using examples from the immune system. We will then discuss the interesting connections between DNA methylation and ageing, and conclude with emerging hypotheses on how this might influence ageing of the immune system.

3.2   DNA Methylation

DNA methylation is found primarily at cytosine bases adjacent to guanine bases, referred to as a cytosine-phosphate-guanine dinucleotides (CpGs). Non-CpG DNA methylation has also been observed in some tissues, but will not be discussed here in detail [2]. The process of DNA methylation is catalyzed by a family of highly conserved enzymes called DNA methyltransferases (DNMTs). In mammals, there are three principal DNMTs: the *de novo* methyltransferases, DNMT3A and DNMT3B, add methyl groups to unmethylated DNA, while DNMT1 is responsible for the maintenance of DNA methylation at hemi-methylated sites during cell division [3]. Until recently, the removal of DNA methylation marks was highly puzzling, as there was evidence for active demethylation, but the enzymes and process responsible remained elusive. In the past few years, new discoveries have shown that DNA demethylation is primarily controlled by 10–11 translocation methylcytosine dioxygenase (TET) enzymes, which catalyze the first step in active demethylation through the base excision repair system [1, 3, 4].

DNA methylation is highly genomic region-specific, partly because CpGs themselves are non-randomly distributed across the genome. CpGs are globally under-represented in genomic sequences, but are enriched in regions called CpG islands: sequences that possess higher than average CpG density and are generally lowly methylated. Interestingly, CpGs are also very dense at many repetitive elements, therefore such high density regions tend to be methylated and silenced to avoid transposition [5, 6]. Nearly 70 % of gene promoters are near a CpG island, and this proximity results in an important relationship between promoter DNA methylation and gene expression [7]. Early models of this relationship posited that high levels of DNA methylation at promoter-associated CpG islands were associated with low levels of expression, and vice versa [8–10] (Fig. 3.1). However, this model is relatively simplistic and we are now beginning to understand that the relationship is much more complex [11–13]. For instance, reverse causation is possible and gene expression levels can influence levels of DNA methylation [12, 14]. In addition, DNA methylation marks in genomic locations other than promoters have equally importance and, in some cases, very different associations with gene expression [12]. For example, both regions of low CpG density directly adjacent to CpG islands, known as "island shores", and enhancer regions do not follow the high DNA methylation/low gene expression model. In island shores, higher levels of DNA methylation are often associated with high levels of gene expression, while at enhancers these relationships have been inconsistent and thus far are poorly understood [1, 5–7, 14, 15].

The underlying genetic sequence can also influence both gene expression and DNA methylation. Reminiscent of earlier work examining the linkage of allelic variation and gene expression, multiple studies have identified SNP/CpG pairs at which genotype is associated with methylation level [8–10, 12, 16–18]. These CpG sites are referred to as methylation quantitative trait loci (mQTLs), and a number of different mechanisms have been hypothesized for their function. For example, the SNP partner may disrupt genetic elements responsible for creating boundaries between methylated and unmethylated genomic regions [11–13, 19]. Alternatively, sequence changes may alter gene expression levels, which in turn influence DNA methylation levels [14]. Regardless of the mechanism by which this relationship functions, it is very important to consider genetic regulation in the analysis of DNA methylation effects.

Fig. 3.1 DNA methylation patterns at promoter-associated CpG islands and gene bodies in active (*above*) and inactive (*below*) genes. Each gene indicates exons (*black*), introns (*white*), CpG islands (*dark grey*) and shores (*light grey*). Below each gene the drawing indicates where DNA methylation is high or low across the gene

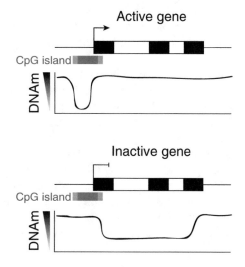

Finally, the relationships between DNA methylation, genotype, and gene expression are dynamic and may change over the life span. For example, it appears that genotype-DNA methylation correlations are susceptible to age-related differences. From early to later life, more than half of the associations between genotype and DNA methylation change [20]. Conversely, correlations between DNA methylation and gene expression appear to be more consistent over the ageing process [21]. Continuing research is required to understand the entire complexity of the role of DNA methylation in genetic regulation across the lifespan.

3.2.1   DNA Methylation and Cell Type Specification

DNA methylation plays a central role in embryogenesis, and this importance continues throughout development [1]. In the immune system, for example, maintenance of DNA methylation patterns is required for hematopoietic stem cell self-renewal and differentiation [22]. During hematopoietic stem cell differentiation, DNA methylation gains and losses at specific regions of the genome "lock-in" differentiation marks that allow cells with the same genetic material to express only the specific genes required for their unique cellular processes and identities [23]. This is illustrated by the lineage-specific epigenetic differences that arise when early multipotent progenitors split into myeloerythroid and lymphoid lineages; these patterns become more specific as differentiation progresses, resulting in cell type specific DNA methylation signatures in mature cells [24–28] (Fig. 3.2). Hematopoietic stem cells lacking the maintenance methyltransferase prematurely lose their self-renewal capabilities, and cells lacking de novo methyltransferases show impaired lineage commitment, illustrating that these DNA methylation changes are essential for lineage development [29, 30].

Because of these cell type specific DNA methylation patterns, cell and tissue types are the largest determinants of DNA methylation variation in healthy individuals [31–33]. It is important to note that these differences in cell type DNA methylation patterns can create at least two major challenges for studies examining the role of DNA methylation in organs or tissues composed of multiple cell types: inter-individual variation in cell types and the concordance between central and surrogate tissues. First, inter-individual cell type differences within a tissue can induce confounding effects that may mask or overwhelm another biological signal with a smaller effect size. In blood, for example, it is essential to control for differences in white blood cell composition between individuals. When cell count information is not available, DNA methylation patterns can be used to predict the underlying cellular composition in order to control for it [28, 34]. This tool is particularly important when studying the relationship between DNA methylation and age in blood tissue, as it has been shown that white blood cell composition changes drastically with age, and that failing to control for these changes can result in white blood cell DNA methylation lineage markers being mistaken for age-associated DNA methylation sites [35]. In regards to tissues other than blood, similar predictive models exist for neurons versus glia in the brain, and other methods exist that can control for cell type differences without specifically predicting underlying cell composition

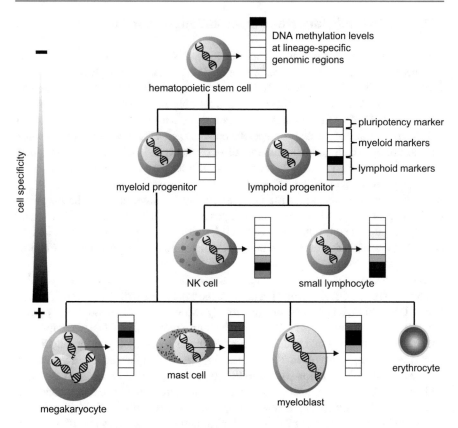

Fig. 3.2 Representation of changes in DNA methylation during hematopoietic stem cell differentiation. Gain and loss of DNA methylation at lineage-specific genomic regions confers cell-type specificity. Methylation patterns become more unique as differentiation progresses. The *shaded boxes* represent DNA methylation at specific regions of the genome. For example, the *black-shaded box* in the hematopoietic stem cell represents DNA methylation levels at genes that enable pluripotency. This mark is slowly lost as the cell becomes more differentiated, and marks of specific cell types arise

[36–38]. The second challenge presented by cell type specific DNA methylation patterns arises when studying surrogate tissues. In human studies, tissues of interest are often inaccessible or require invasive collection methods. To address this, easily collected surrogate tissues, such as blood or cheek swabs, are substituted. However, given the tissue specificity of DNA methylation, it can be challenging to make biological interpretations of function in the tissue of interest when using these alternatives. Ongoing research into the concordance of DNA methylation between tissues collected post-mortem, such as brain and blood, is making the study of surrogate tissues increasingly more interpretable and valuable [33, 39, 40].

3.2.2   DNA Methylation and Environmental Exposures

Complementing its role in cell type specification, DNA methylation is also emerging as a mechanism by which cells "remember" past exposures. Although DNA methylation is stable in that it is generally faithfully transmitted from mother to daughter cells, paradoxically it also appears to be malleable in response to exposures and experiences [41–43]. Researchers have examined environmentally-induced changes in DNA methylation in both gene-specific contexts as well as genome-wide changes, such as average methylation across repetitive elements [44, 45]. These changes may be transient and revert back to their original state after the exposure ends, but in some cases they can remain associated long after the exposure has passed [46]. It is currently hypothesized that early life is a particularly sensitive time for the long-term embedding of epigenetic signatures of exposures, but in many cases it is not until later in life that health outcomes associated with these exposures are revealed. As such, the environmental exposures that accumulate as a person ages can leave behind a biological residue that might influence long-term health.

Some specific environmental exposures have been associated with long-lasting DNA methylation signatures that persist after the exposure itself. For example, a number of DNA methylation changes are strongly associated with previous and current cigarette smoke exposure[47, 48]. The cigarette smoke-related DNA methylation change in the promoter of a well-characterized tumor suppressor gene, aryl hydrocarbon receptor repressor (*AHRR*), is currently the best replicated environmentally induced epigenetic alteration. In the *AHRR* gene, changes in both DNA methylation and gene expression have been observed upon exposure to firsthand and secondhand cigarette smoke, as well as prenatal exposure to maternal smoking [47]. The AHRR protein regulates an enzyme responsible for binding nicotine, thus supporting a plausible mechanism for a DNA methylation response to cigarette smoke exposure.

Lifestyle and the environment can have important effects on DNA methylation, which in turn, may influence immune function, as described in two recent studies. One study examined DNA methylation in African populations with recent divergence in habitat and lifestyle. The study found that a population which had recently diverged into two different habitats, forest versus urban, showed distinct DNA methylation differences between the groups, which were enriched for genes involved in immune function [49]. Another study examined adolescents raised in the American Midwest and compared those who spent the first few years of their lives in Eastern European orphanages to those who were born and raised in their biological families in the US. A significant difference in the ratio of CD4+ T to CD8+ T cells was found in their blood. As well, a DNA methylation pattern unique to the adopted children was enriched for genes involved in development, gene regulation, and behaviour [50]. Together, these examples suggest that the immune system may use DNA methylation as a way to adapt to the environment.

An equally intriguing example of exposure related DNA methylation changes is the reported epigenetic connection between early life adversity and regulation of the inflammatory response in later life. One study examining the association between DNA methylation and early life socioeconomic status (SES) found that while DNA methylation was associated with early life SES, this association was only visible after correction for white blood cell type, reinforcing the evidence for an interaction between DNA methylation and blood cell lineages [11]. Other studies have shown alterations in gene expression and DNA methylation of immune-related genes in adulthood that are associated with low socioeconomic status in childhood [51, 52]. Further work will determine whether this biological signature from early life influences the trajectories of immune ageing.

It is possible that the establishment of these DNA methylation patterns in response to environmental exposures serve to predict future phenotype, including immune responses. For example, pre-stimulation DNA methylation differences in leukocytes can predict their cytokine responses when stimulated through the TLR pathway [11]. As the variability in these baseline patterns was representative of the differences in lifetime environmental exposures between the cells, DNA methylation may function both as a memory of past exposures as well as a predictor of future immune response.

3.2.3   DNA Methylation and Disease

The etiologies of complex diseases are particularly challenging to decode as they often have both environmental and genetic contributors. As DNA methylation has a role in the interface between the genome and the environment, it has been investigated as a mediator between environmental exposure and disease manifestation. DNA methylation could be associated with current or future disease, either mechanistically or as a biomarker. In terms of mechanism, DNA methylation could influence the development of disease via moderation of a genetic risk or embedding of a past environmental exposure, for example. Concurrent disease-related changes in DNA methylation or patterns that predict future disease that do not have an identified mechanistic link to the disease may be biomarkers that could be used for patient stratification or assessment of intervention. Here, examples from the immune system illustrate some of the recent findings connecting DNA methylation and disease.

A number of immune diseases have recently been investigated for epigenetic mediation of genetic risk, in which DNA methylation of immune-related genes alters the penetrance of a risk allele. Investigations of rheumatoid arthritis and peanut allergy have both shown evidence that DNA methylation of the major histocompatibility complex (MHC) in combination with genetic factors alters risk of disease development [53, 54]. The source of the variation in DNA methylation is not known for either of these diseases, but possibilities include stochastic changes or embedded environmental signatures. Future work on the MHC region will be required to elucidate whether environmental exposures alter DNA methylation in that area, as well as whether these can be mitigated to alter disease trajectory.

One of the first autoimmune diseases predicted to have an epigenetic contribution was systemic lupus erythematosus. Potential environmental contributions have led researchers to examine DNA methylation patterns to determine whether genes involved in the production of anti-nuclear antibodies are differentially methylated in lupus patients [55]. In a study examining multiple white blood cell types, abnormally low DNA methylation was observed near interferon-related genes, in both active and quiescent lupus [56]. This implies that the DNA methylation changes are not associated with disease flare-ups, but rather reflect a basal difference in DNA methylation in patients with lupus regardless of their disease status, which may be indicative of molecular "poising" awaiting a specific trigger. The clinical implications of these findings include potential targets for interventions to reduce the active lupus symptoms, or as biomarkers of those who may be at risk for a flare-up.

Other immune diseases show specific patterns of DNA methylation dysregulation, but the mechanistic contribution of DNA methylation is currently not understood. Differences in DNA methylation and gene expression in leukocytes have recently been found between twins discordant for two autoimmune diseases which show generally low concordance: psoriasis and Type 1 diabetes [57, 58]. DNA methylation dysregulation has also been observed in synovial cells of patients with rheumatoid arthritis [59].

DNA methylation has also been identified as a factor involved in microbial infection with two distinct relationships; in some cases it is associated with immune adaptation and in others it is associated with facilitating intracellular parasitic infection. For example, epigenetic remodeling was observed in monocyte-derived dendritic cells in response to *Mycobacterium tuberculosis*, the primary cause of tuberculosis in humans. Infected cells showed DNA methylation changes that altered the regulation of immune transcription factors in a way that contributed to a short-term memory of infection in innate immune cells [60]. Another study demonstrated numerous DNA methylation changes in a macrophage cell line were associated with *Leishmania donovani* infection. Many of these changes were associated with host immune defense pathways such as the AK/STAT signaling pathway and the MAPK signaling pathway [61]. On the other hand, *Anaplasma phagocytophilum*, a prokaryotic pathogen, has been shown to elicit epigenetic changes in human neutrophils that promote infection by enabling pathogen survival and replication [62]. These results highlight an interesting branch of immune epigenetics, showcasing the complex roles DNA methylation has in the process of microbial infection.

In all of these cases, it is not yet known whether the DNA methylation changes observed are a mechanism of disease development or a concurrent biomarker of disease. Regardless of the origin of these changes, they raise the tantalizing possibility that epigenetics in general and DNA methylation in particular may be a new frontier for research into the development of complex immune diseases and phenotypes.

3.3   DNA Methylation and Ageing

Aside from the aforementioned associations of DNA methylation, recent work has shown that DNA methylation also exhibits strong correlations with age, the function of which is not yet known. Since genetic sequence accounts for less than 30 % of lifespan variation, the major driver of human longevity must be attributed to non-genetic factors such as diet, physical activity, smoking and other exposures [63, 64].

Alterations in DNA methylation may occur as a result of the combination of programmed changes in cell type or function with age, the embedding of lifelong environmental exposures, and possible stochastic events over time. Given the combination of evidence that DNA methylation is involved in the embedding of past environments and the development of disease, and that the human methylome is age-sensitive, it seems likely that DNA methylation may have a functional role in ageing. An emerging possible functional role of these age-related epigenetic changes may, in part, play into the molecular pathway responsible for the decline in immune dysfunction observed with age [65]. Over the past years, work has focused on determining the relative contribution of specific environments and stochastic changes in altering DNA methylation patterns during ageing.

3.3.1   Genome-Wide DNA Methylation Ageing Trends

While the link between DNA methylation and ageing has been studied over a long period of time, recent findings and advances have sharpened the focus on the role of DNA methylation in the ageing process. Early studies began assessing the relationship of age and DNA methylation patterns over 40 years ago, using techniques such as liquid chromatography to assess bulk mean methylation levels in salmon, rodent, cattle, and chicken [66, 67]. A pioneering study showed a significant loss of total DNA methylation over the rodent life course across a number of tissues, a finding that was later validated in blood from a cross sectional human cohort consisting of both newborns and centenarians [68, 69]. These explorations laid the foundation for human DNA methylation ageing studies, and as technology continues to advance, allowing easier access to the entire methylome, we enter an exciting era of epigenetic ageing research.

It is estimated that one-third of the epigenome's DNA methylation content changes in association with the ageing process, and recent advances have helped further elucidate the context and potential function of these changes. For example, the previously mentioned finding that DNA exhibits a gradual loss of mean methylation over time has recently been shown to occur in a genomic context-specific manner. Loss of methylation preferentially occurs at regions of low CpG density, often located within a gene body [69]. Despite the fact that mean DNA methylation decreases with age, there are specific age-related methylation changes that involve a gain in methylation as well. These tend to be found within CpG islands, or areas of high CpG density [70, 71]. Together, these changes demonstrate a regression to the mean pattern—low CpG density regions, which are normally highly methylated lose DNA methylation with age, while high density regions which tend to have low levels of DNA methylation gain DNA methylation with age. Since most CpGs in the genome are methylated, this translates to a global loss of DNA methylation. An interesting exception to this pattern is repetitive elements, which tend to be highly methylated and lose DNA methylation with age, despite their high CpG density [72].

Clearly, the relationship between DNA methylation and age is highly complex, with specific patterns occurring at unique genetic regions. A host of research explorations including animal models, human longitudinal twin studies and age-variable

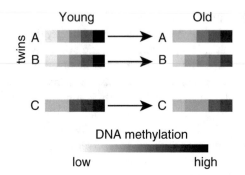

Fig. 3.3 Epigenetic drift results in divergent DNA methylation patterns with increasing age. In early life (*left*), identical twins have highly similar epigenetic patterns (individuals A and B), while individual C is distinct. Later in life (*right*), all individuals are more discordant, as epigenetic drift has altered lowly-methylated CpGs (lighter) to be more methylated, and higher-methylated CpGs (*darker*) to be less methylated

cohorts, have all contributed to identifying DNA methylation patterns with age. From the combination of these research findings, it is evident that two common trends of epigenetic aging have emerged: (1) random changes to DNA methylation that are inconsistent across individuals, and (2) predictable, site-specific DNA methylation changes occurring in a similar way across individuals with age [73, 74] (Fig. 3.3).

3.3.2 Epigenetic Drift

One repeatedly observed feature of age-associated DNA methylation is an increase in variability, resulting from changes in DNA methylation that do not share a common direction (gain or loss) across individuals. These age-related changes are collectively referred to as epigenetic drift, a trend composed of non-directional DNA methylation changes that may be due to stochastic or environmental factors [73–75]. Epigenetic drift accounts for the increasing degree of inter-individual variation across the DNA methylome that occurs with age [76]. This was well-illustrated in an early study of monozygotic twins, which found that at infancy twin pairs possessed almost indistinguishable methylation profiles, while older twin pairs had highly divergent methylomes [77]. The increased inter-individual variability of DNA methylation with age is also reflected in increased variability in transcriptional regulation [78]. Epigenetic drift may not be benign, as variability in DNA methylation has been suggested to increase the risk for diseases such as depression [79]. The source of this increase in variability is still unknown and may represent random DNA methylation events, or an age-associated decline in efficiency of the machinery responsible for maintaining DNA methylation [78, 80]. Others have proposed that an individual's unique combination of lifelong environments and experiences may create differences

Fig. 3.4 Schematic representation of DNA methylation-derived age estimate vs. chronological age. Individuals with a higher DNA methylation age appear biologically older, whereas those with a lower epigenetically approximated age appear biologically younger

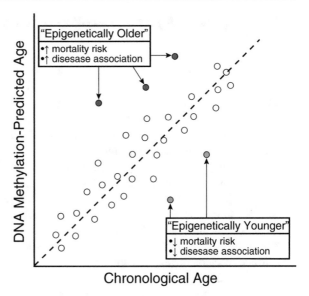

in cellular processes that in turn lead to higher variability in DNA methylation over time [81]. Regardless of its cause, the phenomenon of epigenetic drift is interesting, as it may lead to differential cellular functioning and diverse health outcomes possibly reflected in varying ageing rates.

3.3.3 Epigenetic Clock

In contrast to epigenetic drift, there are CpGs in the genome that are highly associated with age across individuals throughout the life course. These sites follow the same age-related trajectories as epigenetic drift, in which sites that are lowly methylated gain DNA methylation and sites that are highly methylated lose DNA methylation with age [78, 82, 83]. The major difference is that the specific sites that change with age and the direction of that change is consistent across individuals [73]. These sites are in current investigation to determine why they show this consistent relationship, and recently have been used to construct multivariate age predictors, giving rise to the concept of the "epigenetic clock" [78, 82–85] (Fig. 3.4). Several epigenetic clocks have been created to predict biological age within a specific tissue or even across multiple tissues [36, 78, 83, 85]. The current most commonly used age-predictor analyzed over 8000 samples from 51 different cell types to identify 353 sites capable of predicting age with a mean error of 3.6 years [85]. The high correlation between estimated epigenetic age and chronological age supports DNA methylation as a strong biological age predictor.

3.4   Epigenetic Age: A Molecular Marker of Biological Wellbeing?

The widespread application of epigenetic age prediction has shown very high concordance between chronological age and predicted age; however, some individuals show large discrepancies between the two. These efforts have sparked a profusion of studies focused on determining the relationship between lifelong environmental exposures, biological age as measured by the epigenetic clock, and the presence of health and disease during aging.

Recent findings have shown epigenetic age acceleration in a number of diseases and disorders, though few studies have been able to determine whether this acceleration preceded, was concurrent with, or followed disease manifestation in late-onset diseases. For example, neurodegenerative disorders, such as a decline in cognitive function, episodic memory, and working memory, as well as neuropathological measures, such as diffuse and neurotic plaques and amyloid load have been associated with epigenetic age acceleration [86]. In addition, individuals with Down Syndrome, which has been associated with early cognitive decline, have an average epigenetic age 6.6 years older than their chronological age [87]. There have been many other studies showcasing deviations in the relationship between epigenetic age and chronological age in diseases such as Schizophrenia, PTSD, Parkinson's Disease, and HIV [88–92]. In one case, however, researchers were able to show an association between lung cancer incidence and increased epigenetic age acceleration prior to diagnosis [93]. Together, these studies show there are particular diseases or disorders that associate with increased biological age, a relationship consistent with the toll diseases take on human health.

The connection between accelerated epigenetic age and poor health is further reinforced by work analyzing the association between epigenetic age acceleration and all-cause mortality. A longitudinal study found that an epigenetic age more than 5 years older than one's chronological age was associated with a 21% increased mortality rate [94]. The heritability of age acceleration, the degree to which is attributed to genetic composition, was also assessed in a parent-offspring cohort and revealed that approximately 40% of the variation in age acceleration is due to genetic factors [94]. These results show that although a significant proportion of age-related methylation changes may be under a strong genetic influence, there is an even larger unknown non-genetic contribution to the variation in these events. These findings provided one of the first links between DNA methylation-predicted age and mortality, highlighting the potential clinical relevance of age-related DNA methylation.

More recently, another study investigated associations between epigenetic age and mortality in a cohort of 378 Danish twins, aged 30–82 years old. Upon resampling the 86 oldest twins in a 10-year follow-up, a mean 35% higher mortality risk was associated with each 5-year increase in epigenetic age. Interestingly, through a separate intra-pair twin analysis, a 3.2 times greater risk for mortality per 5-year epigenetic age difference within twin pairs was observed for the epigenetically older twin, after controlling for familial factors [95]. This highlights, again, the link between mortality and DNA methylation-predicted age, exemplifying the capacity of DNA methylation to discriminate between biologically younger or older individuals independent of genetic sequence.

The described relationships, where the presence of disease is associated with acceleration in DNA methylation age, which in turn is associated with mortality, are highly suggestive that epigenetic age may be an excellent biomarker of human health. Future work will determine whether acceleration in biological ageing is reversible, and what factors might be involved in modifying the progression of ageing.

3.5   Concluding Remarks

The research highlighted throughout this chapter has only briefly summarized an exceptional amount of work that has contributed to our understanding of the roles DNA methylation plays in health and ageing. It is becoming increasingly evident that DNA methylation functions as the interface of genetic sequence, environmental exposures, and phenotypic outcomes. Thus, it is tempting to speculate that this critical epigenetic mark can be extremely useful to our understanding of health, either through functional contributions or as biomarkers of human phenotypes and diseases.

Although there is an indisputable association between DNA methylation and chronological age throughout the entirety of life, even to the point of developing extremely accurate DNA methylation-based age predictors, it is still not clear what functional role DNA methylation plays in the ageing process. One possible scenario is that epigenetic drift may represent embedding of unique environmental exposures across the lifetime, resulting in increased divergence of DNA methylation profiles with age. Under this model, sites that correlate linearly with chronological age, such as those used for epigenetic clocks, may represent markers of biological ageing and give a molecular insight into the ageing process. On the other hand, it is also possible that age-associated DNA methylation changes do not have a functional component. Under this assumption, epigenetic drift may not reflect environments, but random changes with time, and the epigenetic clock may simply represent those regions of the genome which are more susceptible to age-related changes across individuals [96]. However, the finding that epigenetic age acceleration is correlated with mortality and disease suggests a specific role for common epigenetic changes with age, and further research is needed to clarify its functional relevance.

Despite our gaps in understanding the mechanistic function of DNA methylation in immunity and ageing, it is apparent that methylation is associated with events and exposures that shape an individual's lifelong health. The further we explore these associations, the more it becomes clear that DNA methylation has a significant position in the complex process of ageing and age-related diseases.

References

1. Jones PA. Functions of DNA methylation: islands, start sites, gene bodies and beyond. Nat Rev Genet. 2012;13(7):484–92.
2. Ziller MJ, Müller F, Liao J, Zhang Y, Gu H, Bock C, et al. Genomic distribution and inter-sample variation of non-CpG methylation across human cell types. PLoS Genet. 2011;7(12), e1002389.

3. Smith ZD, Meissner A. DNA methylation: roles in mammalian development. Nat Rev Genet. 2013;14(3):204–20.
4. He Y-F, Li B-Z, Li Z, Liu P, Wang Y, Tang Q, et al. Tet-mediated formation of 5-carboxylcytosine and its excision by TDG in mammalian DNA. Science. 2011;333(6047):1303–7.
5. Kochanek S, Renz D, Doerfler W. DNA methylation in the Alu sequences of diploid and haploid primary human cells. EMBO J. 1993;12(3):1141–51.
6. Alves G, Tatro A, Fanning T. Differential methylation of human LINE-1 retrotransposons in malignant cells. Gene. 1996;176(1-2):39–44.
7. Weber M, Hellmann I, Stadler MB, Ramos L, Pääbo S, Rebhan M, et al. Distribution, silencing potential and evolutionary impact of promoter DNA methylation in the human genome. Nat Genet. 2007;39(4):457–66.
8. Kass SU, Landsberger N, Wolffe AP. DNA methylation directs a time dependent repression of transcription initiation. Curr Biol. 1997;7(3):157–65.
9. Jones PA. The DNA, methylation paradox. Trends Genet. 1999;15(1):34–7.
10. Bell JT, Pai AA, Pickrell JK, Gaffney DJ, Pique-Regi R, Degner JF, et al. DNA methylation patterns associate with genetic and gene expression variation in HapMap cell lines. Genome Biol. 2011;12(1):R10.
11. Lam LL, Emberly E, Fraser HB, Neumann SM, Chen E, Miller GE, et al. Factors underlying variable DNA methylation in a human community cohort. Proc Natl Acad Sci U S A. 2012;109 Suppl 2:17253–60.
12. Gutierrez Arcelus M, Lappalainen T, Montgomery SB, Buil A, Ongen H, Yurovsky A, et al. Passive and active DNA methylation and the interplay with genetic variation in gene regulation. Elife. 2013;2, e00523.
13. Jones MJ, Fejes AP, Kobor MS. DNA methylation, genotype and gene expression: who is driving and who is along for the ride? Genome Biol. 2013;14(7):126.
14. Wagner JR, Busche S, Ge B, Kwan T, Pastinen T, Blanchette M. The relationship between DNA methylation, genetic and expression inter-individual variation in untransformed human fibroblasts. Genome Biol. 2014;15(2):R37.
15. Edgar R, Tan PPC, Portales-Casamar E, Pavlidis P. Meta-analysis of human methylomes reveals stably methylated sequences surrounding CpG islands associated with high gene expression. Epigenetics Chromatin. 2014;7(1):28.
16. Zhang H, Wang F, Kranzler HR, Yang C, Xu H, Wang Z, et al. Identification of methylation quantitative trait loci (mQTLs) influencing promoter DNA methylation of alcohol dependence risk genes. Hum Genet. 2014;133(9):1093–104.
17. Banovich NE, Lan X, McVicker G, van de Geijn B, Degner JF, Blischak JD, et al. Methylation QTLs are associated with coordinated changes in transcription factor binding, histone modifications, and gene expression levels. PLoS Genet. 2014;10(9), e1004663.
18. Fraser HB, Lam LL, Neumann SM, Kobor MS. Population-specificity of human DNA methylation. Genome Biol. 2012;13(2):R8.
19. Maurano MT, Wang H, John S, Shafer A, Canfield T, Lee K, et al. Role of DNA methylation in modulating transcription factor occupancy. Cell Rep. 2015;12(7):1184–95.
20. Smith AK, Kilaru V, Kocak M, Almli LM, Mercer KB, Ressler KJ, et al. Methylation quantitative trait loci (meQTLs) are consistently detected across ancestry, developmental stage, and tissue type. BMC Genomics. 2014;15:145.
21. Peters MJ, Joehanes R, Pilling LC, Schurmann C, Conneely KN, Powell J, et al. The transcriptional landscape of age in human peripheral blood. Nat Commun. 2015;6:8570.
22. Trowbridge JJ, Snow JW, Kim J, Orkin SH. DNA methyltransferase 1 is essential for and uniquely regulates hematopoietic stem and progenitor cells. Cell Stem Cell. 2009;5(4):442–9.
23. Meissner A. Epigenetic modifications in pluripotent and differentiated cells. Nat Biotechnol. 2010;28(10):1079–88. doi:10.1038/nbt.1684. http://www.nature.com/doifinder.

24. Calvanese V, Fernandez AF, Urdinguio RG, Suarez-Alvarez B, Mangas C, Pérez-García V, et al. A promoter DNA demethylation landscape of human hematopoietic differentiation. Nucleic Acids Res. 2012;40(1):116–31.
25. Suarez-Alvarez B, Rodriguez RM, Fraga MF, López-Larrea C. DNA methylation: a promising landscape for immune system-related diseases. Trends Genet. 2012;28(10):506–14.
26. Álvarez-Errico D, Vento-Tormo R, Sieweke M, Ballestar E. Epigenetic control of myeloid cell differentiation, identity and function. Nat Rev Immunol. 2014;15(1):7–17.
27. Reinius LE, Acevedo N, Joerink M, Pershagen G, Dahlen S-E, Greco D, et al. Differential DNA methylation in purified human blood cells: implications for cell lineage and studies on disease susceptibility. PLoS One. 2012;7(7), e41361.
28. Houseman EA, Accomando WP, Koestler DC, Christensen BC, Marsit CJ, Nelson HH, et al. DNA methylation arrays as surrogate measures of cell mixture distribution. BMC Bioinformatics. 2012;13:86.
29. Challen GA, Sun D, Mayle A, Jeong M, Luo M, Rodriguez B, et al. Dnmt3a and Dnmt3b have overlapping and distinct functions in hematopoietic stem cells. Cell Stem Cell. 2014;15(3):350–64.
30. Bröske A-M, Vockentanz L, Kharazi S, Huska MR, Mancini E, Scheller M, et al. DNA methylation protects hematopoietic stem cell multipotency from myeloerythroid restriction. Nat Genet. 2009;41(11):1207–15.
31. Davies MN, Volta M, Pidsley R, Lunnon K, Dixit A, Lovestone S, et al. Functional annotation of the human brain methylome identifies tissue-specific epigenetic variation across brain and blood. Genome Biol. 2012;13(6):R43.
32. Ziller MJ, Gu H, Müller F, Donaghey J, Tsai LTY, Kohlbacher O, et al. Charting a dynamic DNA methylation landscape of the human genome. Nature. 2013;500(7463):477–81.
33. Farré P, Jones MJ, Meaney MJ, Emberly E, Turecki G, Kobor MS. Concordant and discordant DNA methylation signatures of aging in human blood and brain. Epigenetics Chromatin. 2015;8:19.
34. Koestler DC, Christensen B, Karagas MR, Marsit CJ, Langevin SM, Kelsey KT, et al. Blood-based profiles of DNA methylation predict the underlying distribution of cell types: a validation analysis. Epigenetics. 2013;8(8):816–26.
35. Jaffe AE, Irizarry RA. Accounting for cellular heterogeneity is critical in epigenome-wide association studies. Genome Biol. 2014;15(2):R31.
36. Guintivano J, Aryee MJ, Kaminsky ZA. A cell epigenotype specific model for the correction of brain cellular heterogeneity bias and its application to age, brain region and major depression. Epigenetics. 2013;8(3):290–302.
37. Houseman EA, Molitor J, Marsit CJ. Reference-free cell mixture adjustments in analysis of DNA methylation data. Bioinformatics. 2014;30(10):1431–9.
38. Zou J, Lippert C, Heckerman D, Aryee M, Listgarten J. Epigenome-wide association studies without the need for cell-type composition. Nat Methods. 2014;11(3):309–11.
39. Cai C, Langfelder P, Fuller TF, Oldham MC, Luo R, van den Berg LH, et al. Is human blood a good surrogate for brain tissue in transcriptional studies? BMC Genomics. 2010;11(1):589.
40. Hannon E, Lunnon K, Schalkwyk L, Mill J. Interindividual methylomic variation across blood, cortex, and cerebellum: implications for epigenetic studies of neurological and neuro-psychiatric phenotypes. Epigenetics. 2015;10(11):1024–32.
41. Marsit CJ. Influence of environmental exposure on human epigenetic regulation. J Exp Biol. 2015;218(Pt 1):71–9.
42. Boyce WT, Kobor MS. Development and the epigenome: the "synapse" of gene-environment interplay. Dev Sci. 2015;18(1):1–23.
43. Feil R, Fraga MF. Epigenetics and the environment: emerging patterns and implications. Nat Rev Genet. 2012;13(2):97–109.
44. Baccarelli A, Tarantini L, Wright RO, Bollati V, Litonjua AA, Zanobetti A, et al. Repetitive element DNA methylation and circulating endothelial and inflammation markers in the VA normative aging study. Epigenetics. 2010;5(3):222–8.
45. Bollati V, Baccarelli A. Environmental epigenetics. Heredity. 2010;105(1):105–12.

46. Essex MJ, Boyce WT, Hertzman C, Lam LL, Armstrong JM, Neumann SMA, et al. Epigenetic vestiges of early developmental adversity: childhood stress exposure and DNA methylation in adolescence. Child Dev. 2013;84(1):58–75.
47. Monick MM, Beach SRH, Plume J, Sears R, Gerrard M, Brody GH, et al. Coordinated changes in AHRR methylation in lymphoblasts and pulmonary macrophages from smokers. Am J Med Genet B Neuropsychiatr Genet. 2012;159B(2):141–51.
48. Shenker NS, Polidoro S, van Veldhoven K, Sacerdote C, Ricceri F, Birrell MA, et al. Epigenome-wide association study in the European Prospective Investigation into Cancer and Nutrition (EPIC-Turin) identifies novel genetic loci associated with smoking. Hum Mol Genet. 2013;22(5):843–51.
49. Fagny M, Patin E, MacIsaac JL, Rotival M, Flutre T, Jones MJ, et al. The epigenomic landscape of African rainforest hunter-gatherers and farmers. Nat Commun. 2015;6:10047.
50. Esposito EA, Jones MJ, Doom JR, MacIsaac JL, Gunnar MR, Kobor MS. Differential DNA methylation in peripheral blood mononuclear cells in adolescents exposed to significant early but not later childhood adversity. Dev Psychopathol. 2016:1–15.
51. Stringhini S, Polidoro S, Sacerdote C, Kelly RS, van Veldhoven K, Agnoli C, et al. Life-course socioeconomic status and DNA methylation of genes regulating inflammation. Int J Epidemiol. 2015;44(4):1320–30.
52. Miller GE, Chen E, Fok AK, Walker H, Lim A, Nicholls EF, et al. Low early-life social class leaves a biological residue manifested by decreased glucocorticoid and increased proinflammatory signaling. Proc Natl Acad Sci U S A. 2009;106(34):14716–21.
53. Hong X, Hao K, Ladd-Acosta C, Hansen KD, Tsai H-J, Liu X, et al. Genome-wide association study identifies peanut allergy-specific loci and evidence of epigenetic mediation in US children. Nat Commun. 2015;6:6304.
54. Liu Y, Aryee MJ, Padyukov L, Fallin MD, Hesselberg E, Runarsson A, et al. Epigenome-wide association data implicate DNA methylation as an intermediary of genetic risk in rheumatoid arthritis. Nat Biotechnol. 2013;31(2):142–7.
55. Zhang Y, Zhao M, Sawalha AH, Richardson B, Lu Q. Impaired DNA methylation and its mechanisms in CD4(+)T cells of systemic lupus erythematosus. J Autoimmun. 2013;41:92–9.
56. Absher DM, Li X, Waite LL, Gibson A, Roberts K, Edberg J, et al. Genome-wide DNA methylation analysis of systemic lupus erythematosus reveals persistent hypomethylation of interferon genes and compositional changes to CD4+ T-cell populations. PLoS Genet. 2013;9(8), e1003678.
57. Gervin K, Vigeland MD, Mattingsdal M, Hammerø M, Nygård H, Olsen AO, et al. DNA methylation and gene expression changes in monozygotic twins discordant for psoriasis: identification of epigenetically dysregulated genes. PLoS Genet. 2012;8(1), e1002454.
58. Stefan M, Zhang W, Concepcion E, Yi Z, Tomer Y. DNA methylation profiles in type 1 diabetes twins point to strong epigenetic effects on etiology. J Autoimmun. 2014;50:33–7.
59. Laricade L, Urquiza JM, Gomez-Cabrero D, Islam ABMMK, López-Bigas N, Tegnér J, et al. Identification of novel markers in rheumatoid arthritis through integrated analysis of DNA methylation and microRNA expression. J Autoimmun. 2013;41:6–16.
60. Pacis A, Tailleux L, Morin AM, Lambourne J, MacIsaac JL, Yotova V, et al. Bacterial infection remodels the DNA methylation landscape of human dendritic cells. Genome Res. 2015;25(12):1801–11.
61. Marr AK, MacIsaac JL, Jiang R, Airo AM, Kobor MS, McMaster WR. Leishmania donovani infection causes distinct epigenetic DNA methylation changes in host macrophages. PLoS Pathog. 2014;10(10), e1004419. http://dx.plos.org/10.1371/journal.ppat.1004419.
62. Sinclair SHG, Yegnasubramanian S, Dumler JS. Global DNA methylation changes and differential gene expression in Anaplasma phagocytophilum-infected human neutrophils. Clin Epigenetics. 2015;7(1):77.
63. Gebel K, Ding D, Chey T, Stamatakis E, Brown WJ, Bauman AE. Effect of moderate to vigorous physical activity on all-cause mortality in middle-aged and older Australians. Am Medical Assoc. 2015;175(6):970–7.

64. Christensen K, Johnson TE, Vaupel JW. The quest for genetic determinants of human longevity: challenges and insights. Nat Rev Genet. 2006;7(6):436–48.
65. Goronzy JJ, Li G, Weyand CM. DNA methylation, age-related immune defects, and autoimmunity. Epigenetics of aging. New York: Springer; 2010. p. 327–44.
66. Romanov GA, Vanyushin BF. Methylation of reiterated sequences in mammalian DNAs. Effects of the tissue type, age, malignancy and hormonal induction. Biochim Biophys Acta. 1981;653(2):204–18.
67. Hoal-van Helden EG, van Helden PD. Age-related methylation changes in DNA may reflect the proliferative potential of organs. Mutat Res. 1989;219(5-6):263–6.
68. Wilson VL, Smith RA, Ma S, Cutler RG. Genomic 5-methyldeoxycytidine decreases with age. J Biol Chem. 1987;262(21):9948–51.
69. Heyn H, Li N, Ferreira HJ, Moran S, Pisano DG, Gomez A, et al. Distinct DNA methylomes of newborns and centenarians. Proc Natl Acad Sci U S A. 2012;109(26):10522–7.
70. Johansson A, Enroth S, Gyllensten U. Continuous aging of the human DNA methylome throughout the human lifespan. PLoS One. 2013;8(6), e67378.
71. Christensen BC, Houseman EA, Marsit CJ, Zheng S, Wrensch MR, Wiemels JL, et al. Aging and environmental exposures alter tissue-specific DNA methylation dependent upon CpG island context. PLoS Genet. 2009;5(8), e1000602.
72. Bollati V, Schwartz J, Wright R, Litonjua A, Tarantini L, Suh H, et al. Decline in genomic DNA methylation through aging in a cohort of elderly subjects. Mech Ageing Dev. 2009;130(4):234–9.
73. Jones MJ, Goodman SJ, Kobor MS. DNA methylation and healthy human aging. Aging Cell. 2015;14(6):924–32.
74. Boks MP, van Mierlo HC, Rutten BPF, Radstake TRDJ, De Witte L, Geuze E, et al. Longitudinal changes of telomere length and epigenetic age related to traumatic stress and post-traumatic stress disorder. Psychoneuroendocrinology. 2015;51:506–12.
75. Teschendorff AE, West J, Beck S. Age-associated epigenetic drift: implications, and a case of epigenetic thrift? Hum Mol Genet. 2013;22(R1):R7–15.
76. Martin GM. Epigenetic drift in aging identical twins. Proc Natl Acad Sci U S A. 2005;102(30):10413–4.
77. Fraga MF, Ballestar E, Paz MF, Ropero S, Setien F, Ballestart ML, et al. Epigenetic differences arise during the lifetime of monozygotic twins. Proc Natl Acad Sci U S A. 2005;102(30):10604–9.
78. Hannum G, Guinney J, Zhao L, Zhang L, Hughes G, Sadda S, et al. Genome-wide methylation profiles reveal quantitative views of human aging rates. Mol Cell. 2013;49(2):359–67.
79. Córdova-Palomera A, Fatjó-Vilas M, Gastó C, Navarro V, Krebs M-O, Fañanás L. Genome-wide methylation study on depression: differential methylation and variable methylation in monozygotic twins. Transl Psychiatry. 2015;5, e557.
80. Winnefeld M, Lyko F. The aging epigenome: DNA methylation from the cradle to the grave. Genome Biol. 2012;13(7):165.
81. Murgatroyd C, Patchev AV, Wu Y, Micale V, Bockmühl Y, Fischer D, et al. Dynamic DNA methylation programs persistent adverse effects of early-life stress. Nat Neurosci. 2009;12(12):1559–66.
82. Bocklandt S, Lin W, Sehl ME, Sanchez FJ, Sinsheimer JS, Horvath S, et al. Epigenetic predictor of age. PLoS One. 2011;6(6), e14821.
83. Weidner CI, Lin Q, Koch CM, Eisele L, Beier F, Ziegler P, et al. Aging of blood can be tracked by DNA methylation changes at just three CpG sites. Genome Biol. 2014;15(2):R24.
84. Koch CM, Wagner W. Epigenetic-aging-signature to determine age in different tissues. Aging (Albany NY). 2011;3(10):1018–27.
85. Horvath S. DNA methylation age of human tissues and cell types. Genome Biol. 2013;14(10):R115.
86. Levine ME, Lu AT, Bennett DA, Horvath S. Epigenetic age of the pre-frontal cortex is associated with neuritic plaques, amyloid load, and Alzheimer's disease related cognitive functioning. Aging (Albany NY). 2015;7(12):1198–211.

87. Horvath S, Garagnani P, Bacalini MG, Pirazzini C, Salvioli S, Gentilini D, et al. Accelerated epigenetic aging in Down syndrome. Aging Cell. 2015;14(3):491–5.
88. van Eijk KR, de Jong S, Strengman E, Buizer-Voskamp JE, Kahn RS, Boks MP, et al. Identification of schizophrenia-associated loci by combining DNA methylation and gene expression data from whole blood. Eur J Hum Genet. 2015;23(8):1106–10.
89. Wolf EJ, Logue MW, Hayes JP, Sadeh N, Schichman SA, Stone A, et al. Accelerated DNA methylation age: Associations with PTSD and neural integrity. Psychoneuroendocrinology. 2015;63:155–62.
90. Horvath S, Levine AJ. HIV-1 infection accelerates age according to the epigenetic clock. J Infect Dis. 2015;212(10):1563–73.
91. Rickabaugh TM, Baxter RM, et al. Acceleration of age-associated methylation patterns in HIV-1-infected adults. PLoS One. 2015;10(3), e0119201. http://dx.plos.org/10.1371/journal.pone.0119201.
92. Horvath S, Ritz BR. Increased epigenetic age and granulocyte counts in the blood of Parkinson's disease patients. Aging (Albany NY). 2015;7(12):1130–42.
93. Levine ME, Hosgood HD, Chen B, Absher D, Assimes T, Horvath S. DNA methylation age of blood predicts future onset of lung cancer in the women's health initiative. Aging (Albany NY). 2015;7(9):690–700.
94. Marioni RE, Shah S, McRae AF, Chen BH, Colicino E, Harris SE, et al. DNA methylation age of blood predicts all-cause mortality in later life. Genome Biol. 2015;16(1):25.
95. Christiansen L, Lenart A, Tan Q, Vaupel JW, Aviv A, McGue M, et al. DNA methylation age is associated with mortality in a longitudinal Danish twin study. Aging Cell. 2015;15(1):149–54.
96. Reynolds LM, Taylor JR, Ding J, Lohman K, Johnson C, Siscovick D, et al. Age-related variations in the methylome associated with gene expression in human monocytes and T cells. Nat Commun. 2014;5:5366.

The Role of CMV in Immunosenescence

4

Ludmila Müller, Klaus Hamprecht, and Graham Pawelec

Abstract

The term "immunosenescence" is commonly taken to mean age-associated changes in immune parameters hypothesized to contribute to increased susceptibility and severity of the older adult to infectious disease, autoimmunity and cancer. In humans, it is characterized by lower numbers and frequencies of naïve T and B cells and higher numbers and frequencies of late-differentiated T cells, especially CD8+ T cells, in the peripheral blood. The latter may be very noticeable, but intriguingly, only in people infected by human herpesvirus 5 (Cytomegalovirus, CMV). Almost all human studies have been cross-sectional, thus documenting differences between old and young populations, but not necessarily changes over time. Nonetheless, limited longitudinal studies have provided data consistent with gradually decreasing naïve T and B cells, and increasing late-differentiated T cells over time, and in rare instances associating these changes with increasing frailty and incipient mortality in the elderly. Low numbers of naïve cells render the aged highly susceptible to pathogens to which they have not been previously exposed, but are not otherwise associated with an "immune risk profile" predicting earlier mortality. Whether the accumulations of late-differentiated T cells driven primarily by CMV contribute to frailty and mortality or are only adaptive responses to the persistent virus remains controversial.

L. Müller
Max Planck Institute for Human Development, Lentzeallee 94, 14195 Berlin, Germany
e-mail: lmueller@mpib-berlin.mpg.de

K. Hamprecht
Institute of Medical Virology, University of Tübingen, Tübingen, Germany
e-mail: klaus.hamprecht@med.uni-tuebingen.de

G. Pawelec (✉)
Center for Medical Research, University of Tübingen,
Waldhörnlestr. 22, 72072 Tübingen, Germany

Division of Cancer Studies, Faculty of Life Sciences and Medicine, King's College London, London, UK

The John van Geest Cancer Research Centre, School of Science and Technology, Nottingham Trent University, Nottingham, UK
e-mail: graham.pawelec@uni-tuebingen.de

© Springer International Publishing Switzerland 2017
V. Bueno et al. (eds.), *The Ageing Immune System and Health*,
DOI 10.1007/978-3-319-43365-3_4

Either way, there is currently little direct evidence that "immunosenescence" contributes to either autoimmunity or cancer in the aged. This chapter reviews some of the studies implicating CMV infection in immunosenescence and its consequences for ageing trajectories in humans.

Keywords

Ageing • Immune system • CMV • Immunosenescence • Health • Vaccination

Abbreviations

CCR7	Chemokine receptor 7
CMV	Cytomegalovirus
DC	Dendritic cell
γδ T cells	Gamma-delta T cells
IE-1	Immediate early protein 1
IFN	Interferon
KLRG	Killer lectin-like cell receptor G1
MHC	Major histocompatibility complex
NK cell	Natural killer cell
PBMC	Peripheral blood mononuclear cells
pp65	Tegument protein
TLR	Toll-like receptor

4.1   Introduction

The ageing of the immune system represents a universal, dynamic and multifaceted process characterized by progressive immunodeficiency, chronic inflammation, and autoimmunity [1, 2]. The changes occurring in the immune system with ageing are commonly collectively designated "immunosenescence" but this term should be reserved for parameters shown to be detrimental, according to the meaning of the word "senescence". It remains true to say that many of the changes, or most often, differences, in immune parameters of the older adult relative to the young have not actually been shown to be detrimental. The realization that compensatory changes may be developing over time is gaining ground [3, 4].

The most noticeable differences between old and young people are observed in the adaptive immune system and especially in T cells (see Chap. 2). A large body of experimental work has been devoted to the investigation of age-related differences in T-cell phenotypes and functions in young and old individuals, but few longitudinal studies in humans actually delineate changes at the level of the individual. Data not only from cross-sectional studies, but also from these few longitudinal studies, have mostly confirmed that the major age-associated differences are seen in the CD8⁺ rather than the CD4⁺ subset.

Several studies have now shown that lower percentages and absolute numbers of naïve CD8$^+$ T cells are seen in all older subjects whereas the accumulation of very large numbers of CD8$^+$ late-stage differentiated memory cells is seen in a majority but not in all older adults [2]. The major difference between this majority of subjects with such accumulations of memory cells and those without is that the former are infected with human herpesvirus 5 (Cytomegalovirus, CMV). Nevertheless, the question of whether CMV is associated with immunosenescence remains so far uncertain as no causal relationship has been unequivocally established [5]. Because changes are seen rapidly after primary infection in transplant patients [6] and infants [7], it is highly likely that CMV does drive the accumulation of CD8$^+$ late-stage memory cells, but the relationship of this to senescence remains unclear.

4.2   Cytomegalovirus Infection, Defense Strategies and Development of Latency

CMV is a double-stranded DNA virus belonging to the β-herpesvirus family and is an omnipresent pathogen worldwide. In developing countries, the percentage of the population infected with CMV early in life approximates 100 %, but in industrialized societies this is far less and depends on subject age and socioeconomic factors [8]. Thus, it may be considered that not being infected with CMV is an artefact of civilization and hygiene. However, the fraction of the infected population in developed countries is commonly higher in the aged than the young, either because of new infections and seroconversion throughout the lifespan [8] and/or because younger and older people experienced different exposures in childhood and later life. This quandary again illustrates a basic problem with studies of this type in humans: they are cross-sectional and therefore it is formally impossible to distinguish between increases over time within individuals, and differences in the circumstances of the populations studied, as in this case. Hence, we must be careful to state "differences" when the data are from cross-sectional studies and limit "changes" to those that we know for certain do occur over time.

The genome of wild-type CMV strains is of about 235 kb with ca. 200 genes; it is the largest known human virus. Its double-stranded DNA is enclosed in an icosahedral viral capsid, which itself is surrounded by tegument or matrix proteins (Fig. 4.1). CMV has a class E genome: the two unique regions, unique long (UL) and unique short (US), are flanked by pairwise inverted repeats (terminal or internal repeat long, TRL/IRL) and internal/terminal repeat short (IRS/TRS): 5′ TRL—IRL-IRS—TRS. The "α sequence" is shared by both repeats. Additional regions include the "b and c sequence". The genome exists in four genomic isomers by inversion of the UL/US regions. In the laboratory, one must discriminate between high (Ad169, Towne) and low passage (Toledo, Merlin) CMV strains: Ad169 and Towne are highly passaged laboratory mutants, which lack either 15 or 13 kb, respectively, at the 3′end of the UL region (UL/b′region). Nearly all strains primarily isolated and propagated in fibroblast culture show mutations in the UL128 locus (UL128L) and in genes of the RL11 family, which has an impact on cell

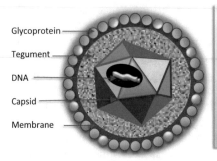

Glycoprotein

Tegument

DNA

Capsid

Membrane

- Belongs to the family of β-herpesviruses
- Lipid membrane is riddled with glycoproteins
- CMV contains an icosahedral protein capsid
- Possesses a 235-kb double-stranded DNA
- Tegument includes most bioactive proteins (TPs)
- Tegument proteins are responsible for viral entry
- TPs regulate viral replication and gene expression
- TPs play major role for immune evasion of CMV
- TPs participate in assembly and egress of new CMV

Fig. 4.1 Basic components of cytomegalovirus

tropism. UL128L includes the UL128, UL130, and UL131A regions, which form a pentameric complex with viral glycoprotein genes gH and gL. This pentameric complex is not essential for viral entry into fibroblasts but it is essential for endothelial and epithelial entry. Additionally, 14 genes located at the 5′ end of the genome (RL5A, RL6, RL11—UL1, UL4-UL11) are not essential for viral growth in fibroblasts [9].

The central part of the UL region contains clusters of genes which have homologs in other herpesviruses, such as DNA polymerase, glycoprotein B and glycoprotein H, while the rest of the genome contains genes that appear primarily only in β-herpes viruses or exclusively in human CMV [10, 11]. Thus, there is great potential complexity in the interactions of CMV and its products with infected cells and the immune system. These are still far from clarified.

CMV is usually naturally acquired as an asymptomatic primary infection in infancy or during the first months of life via breastfeeding, after which the virus establishes lifelong persistence. Transmission of the virus occurs through exposure to infectious body fluids, including saliva, urine, breast milk, semen and blood [12]. Both T cells and antibodies specific for CMV remain present throughout life, although the possibility that some individuals may harbor CMV and specific T cells, but no antibody, has not been completely excluded [10, 13].

The initial infection with CMV activates a whole series of complex reactions of the host in order to limit the replication of the virus (Fig. 4.2). The diverse defense systems, in which the innate and adaptive immune system are involved, are able to detect the foreign nature of the virus very early after contact. Specialized pathogen-associated receptors recognize viral glycoproteins and even the entire viral genome as such. In the early phase of the immune response interferons and other cytokines play an important role in keeping the viral infection in check [11]. Additionally, NK cells as well as T cells and B cells with their specific cytokines and cytotoxic activity and antibody production, all contribute to CMV surveillance [10, 11]. Neutralizing antibodies produced by B cells are thought to play an important role in controlling CMV infection and re-infection. The impaired production of effector B cells secreting antibody specific for envelope glycoprotein and the induction of atypical memory B cells may reduce the production of neutralizing antibodies during primary CMV infection and support virus dissemination and establishment of CMV latency

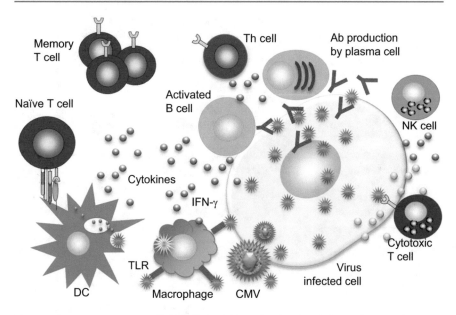

Fig. 4.2 Immune responses to CMV. DC: dendritic cells; Th cell: T helper cell; NK cell: natural killer cell; IFN: interferon; TLR: toll-like receptor; CMV: cytomegalovirus

[14]. As the proportion of the older population infected with CMV is greater than the young, the effects of CMV may be confused with those of chronological age.

Given the numerous host defense mechanisms (Fig. 4.2), successful viral infection is highly dependent on multiple evasive strategies (Fig. 4.3). These include some proteins that are delivered together with infectious virions (such as tegument protein pp65) and are activated during the early phase of infection (such as viral IE proteins) in order to block intrinsic cellular defense mechanisms [15]. CMV is also able to interfere with components of the cellular immune response. For example, products of at least seven CMV genes have been identified that inhibit NK-cell functions [16, 17]. Several other genes modulate the presentation of CMV peptides to T cells. They inhibit the loading of peptides onto MHC complexes or cause a dislocation of class I molecules from endoplasmic reticulum into the cytosol, where they are then degraded [18]. CMV also encodes a number of molecules similar to chemokines, cytokines, and their receptors, which are probably involved in immune evasion [19]. The virus also inhibits the proliferation of peripheral blood mononuclear cells and production of cytokines from these cells [10]. In addition, CMV inhibits the maturation of dendritic cells and is able also to initiate the process of apoptosis [10, 11, 20].

In this way, CMV infection initiates a complex interaction between the virus and host. Despite the sophisticated mechanisms with which CMV is equipped to subvert the innate and adaptive immune responses of the host, the latter are generally able to clear cells producing infectious virus particles in immunocompetent individuals [21, 22]. Also intrinsic defenses, which are cell-autonomous components of the immune system protecting individual cells from viral attack, play an important role

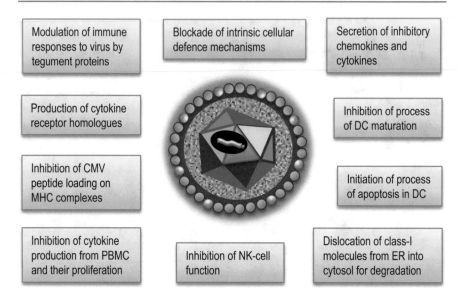

Fig. 4.3 Examples of evasive strategies of CMV. CMV: cytomegalovirus; MHC: major histocompatibility complex; PBMC: peripheral blood mononuclear cells; NK cell: natural killer cell; DC: dendritic cells; ER: endoplasmic reticulum

[23]. They are mediated by constitutively expressed proteins called restriction factors, and can rapidly act at a very early stage of CMV infection by inhibiting an essential viral process. This may be one reason why herpes viruses have evolved to establish a latent state of infection, during which infectious virions are not generated, and the virus is able to persist for the lifetime of the host, but does not cause active disease (Fig. 4.4). Although various cell types may support productive infection with CMV and carry the virus, latent CMV infection has been most convincingly documented in cells of myeloid origin [21]. In the bone marrow, CD34+ myeloid progenitor cells are a site of latency, while in peripheral blood CD14+ monocytes may carry latent CMV. Intense research on the clarification of the molecular basis for cell type-specific CMV latency remains at an early stage. It is still not definitively established how CMV establishes latency after acute infection in the face of an ongoing robust immune response involving both innate and adaptive immunity. The results of some studies indicate that the decision between permissive and latent infection may be determined by the balance between activating and repressive factors, which control transcription of viral genes upon initial infection, and this may differ among different cell types [22, 24].

It is also becoming increasingly clear that the latent state might be mediated by epigenetic factors through the process of heterochromatinization of viral genomes to silence viral gene expression. As a result, the virus can indefinitely persist in these cells because it does not express pathogen-associated molecular patterns, rendering it invisible to the host immune system [25]. Thus, CMV is able to manipulate cellular controls including regulation of the cell cycle and gene silencing extremely

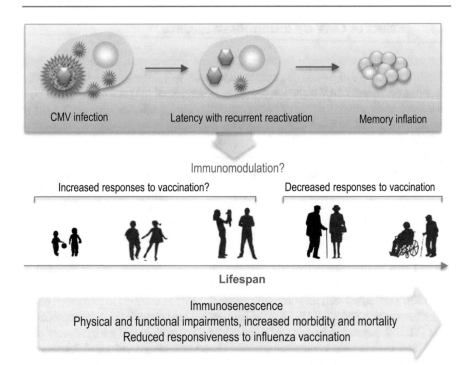

Fig. 4.4 Impact of CMV infection on immunosenescence

well. The ability to overcome cellular defenses as well as to control of proliferation and gene expression are decisive factors, which allow future viral replications and/or viral persistence [5, 21, 22, 25, 26].

In a cell culture model of latency, it was shown that CMV expresses only a restricted number of genes. Among them is an alternatively spliced, latency-associated form of IL-10-like molecule (LAcmvIL-10) that is known to inhibit MHC class II recognition of infected cells, thus avoiding elimination by CD4$^+$ T cells of the adaptive immune system. This immunomodulatory cytokine, secreted from latent virus-infected cells potentially over decades could contribute generally to chronic immune system dysfunction [18]. During experimental latency in CD34$^+$ cells, cellular miRNAs (for example hsa-miR-92a) can be downregulated, leading to upregulation of the cellular myeloid transcription factor GATA2, which may be important for proliferation and survival of hematopoietic progenitor cells. In this model, viral LAcmvIL-10 induces hsa-miR-92a, which upregulates GATA2. This may lead to transcription of latent CMV genes (LUNA, UL144) and also of the cellular cytokine gene IL-10. Finally, IL-10 may block apoptosis of latently infected cells (via BCl-2, STAT3 phosphorylation and HSP-70) [24]. The summed side-effects of lifelong coexistence with CMV are likely to have profound implications for the functioning of the host's immune system and the phenomena of immunosenescence.

4.3   Effect of CMV on Immunosenescence

The cause-effect relationship between CMV and immunosenescence is not clear, despite the plethora of candidate contributory factors mentioned above. It is, however, clear that persistent CMV infection results in expansion of the total CD8$^+$ T-cell pool associated with accumulations of late-stage differentiated effector memory T cells in the periphery. In CMV-seropositive people, especially older people, a remarkably high fraction of circulating CD8$^+$ T lymphocytes is often found to be specific for CMV. However, although the proportion of naïve CD8$^+$ T cells is lower in the old than the young whether or not they are CMV-infected, the gross accumulation of late-stage differentiated CD8$^+$ T cells only occurs in CMV-seropositive individuals [27]. This phenomenon of gradual increase in CMV-specific CD8$^+$ T-cell population over time is known as memory inflation [28, 29]. It is not clear whether this is adaptive or pathological; however, a lower fraction of CD8$^+$ T cells bearing receptors for the HLA-A2-NLV epitope of CMV pp65 is able to produce IFN-γ in response to specific peptide in the old than in the young. Nonetheless, the absolute number of T cells assessed as functional by this method is greater in the old than in the young, due to memory inflation [30]. The large T-cell responses elicited by CMV are ascribed to numerous CMV proteins of which pp65 and IE-1 are the best defined and considered as immunodominant [31, 32]. Remarkably, this CMV-specific, fully functional effector memory pool of CD8$^+$ T cells is able to persist for a long time, suggesting a unique differentiation pathway that apparently differs from the generation of short-lived CD8$^+$ memory T cells generated after infection with acute viruses [33] or even closely-related persistent herpes viruses such as Epstein-Barr virus. The reasons for such differences are unknown but may be related to the type of cell hosting the latent virus.

It is assumed that the latent phase of CMV infection can be accompanied by recurrent episodes of viral reactivation, but this has been hard to document in humans [34]. This is mostly a subclinical process, which leads to production of immunogenic transcripts that may maintain memory inflation of virus-specific cytotoxic lymphocytes. Supporting this idea it was demonstrated, in a murine model, that treatment with valaciclovir (that inhibits CMV replication) leads to a dramatic suppression of MCMV-specific T-cell memory subset and regeneration of the naïve CD8$^+$ T-cell compartment [35]. This could be taken to represent a reversal of a classic marker of immunosenescence. Results from the mouse model also demonstrated that latent CMV infection impairs immunity in old animals [36]. Such experiments have not been carried out on humans, but studies on people receiving anti-HSV drugs which also have some minor effects on T cell responses to CMV pp65 have been published [37]. Results in mice are clear [35] and in humans recent data show that CMV viral load increases after the age of 70, suggesting potential slippage of the strict control of latency [38].

The total CMV-specific T-cell response in seropositive subjects constitutes on average approximately 10 % of both the CD4$^+$ and CD8$^+$ memory compartments, and can be far greater in older people. It has been documented that cross-reactive recognition of CMV proteins in seronegative individuals is rare and is limited exclusively to CD8$^+$ T cells, so that most of the cells responding to CMV are likely to be truly CMV-specific and not cross-reactive [39]. Because of the high level of both CD4$^+$

and CD8+ memory T cells in older people, the idea arose that CMV promotes immunosenescence in a clinically-relevant sense, i.e. associated with mortality [30, 40–42]. This was based on original findings that an inverted CD4:8 cell ratio caused by accumulated CD8+ memory cells specific for CMV was associated with 2, 4 and 6-year mortality of very old people in the longitudinal Swedish OCTO/NONA studies. These studies remain some of the very few indicating detrimental effects of the immune changes observed, meeting the definition of "senescence" as being demonstrably detrimental. The cluster of parameters including lower numbers of B cells as well as T cell parameters, and CMV-seropositivity, was designated an "Immune Risk Profile" (IRP). However, due to lack of other studies of this type, it remains unclear whether the IRP is informative in other populations and whether it has any relevance to younger people not already selected for survival into advanced age. Different risk factors may be relevant in other populations, for example, in the BELFRAIL study of Flemings around 83 years old at baseline. In this study, it was found that a naïve T-cell-dominated CD4:8 ratio >5 rather than the IRP-defining ratio of <1 in the Swedish population, and which was absent in young donors, was associated with a higher physical and functional impairment in very old people infected with CMV [43]. Moreover, further studies have revealed that this profile is associated with 3-year mortality in women, but not men (Adriaensen et al., unpublished observations). Unlike in the Swedish studies an IRP characterized by an inverted CD4:8 ratio was not informative for survival in this Belgian population. This finding suggests that in a very old population not only CMV infection but also the degree of concomitant immune dysregulation play an important role with regard to physical impairment and consequences for health state, and these may well be very sensitive to the early-life and current circumstances of the population studied. In fact, there may be disparities of such major proportions that in some cases at least it could be possible to discern an advantage to being CMV-infected in early life, as reflected in the "hygiene hypothesis". For example, as mentioned above, there are some published data suggesting that that in young humans or young mice, CMV may improve immune responses to some antigens and to influenza virus, probably by way of increased pro-inflammatory responses [44, 45]. There is also an animal model in which CMV protects against fatal infections with certain other pathogenic organisms [46]. These observations suggest that the effect of CMV on the immune system may be highly dependent also on an individuals' age and circumstances, and that what is viewed as ageing is in fact later collateral damage from immune reactivity that was beneficial in earlier life [47, 48]. This is saying nothing more than that the same immune pathology that always accompanies immune responses to acute viruses is also caused by CMV, but over a chronic time scale and usually subclinical. Because the magnitude of the CMV-specific immune response increases with age and accordingly with the chronological history of co-existence with an antigen, this leads to ever-increasing memory inflation and potential collateral damage [30, 35, 40].

Another important variable that appears to drive higher expansions of memory CMV-specific effector T cells was shown to be a higher viral load [49], which, as mentioned above, is also greater in the aged [38]. The influence of the viral inoculum on the degree of memory T-cell inflation may provide an explanation for the observed variations in the magnitude and phenotype of CMV-specific T-cell responses in people [50], only some of which is likely to have a genetic basis [27]. Thus, as we saw above, in general the CMV-specific CD8$^+$ T-cell response is characterized in the periphery by an accumulation of late-stage differentiated effector cells, sometimes referred to in the literature as "terminally" differentiated, which have downregulated expression of the costimulatory molecules CD27 and CD28. This population of cells is characterized by expression of the negative costimulatory molecules KLRG-1 and CD57 and re-expression of CD45RA. They have lost the homing molecules CCR7 and CD62L as well the cytokine receptors CD122 and CD127 [3, 33, 51, 52]. Although some T cells with this phenotype are present in CMV-seronegative people, they are overwhelmingly present in CMV-seropositive adults [27]. They maintain high levels of cytotoxic molecules such as granzyme B but many of them seem not to produce perforins and are therefore compromised in their killing ability. Inappropriate release of superfluous granzyme B has been postulated to cause much of the tissue damage associated with diseases of ageing [53]. Many of these cells retain the ability to produce the pro-inflammatory cytokine IFN-γ upon mitogenic stimulation but the fraction of those doing so upon stimulation with CMV antigens is lower in the elderly than in the young, at least for the one specific CMV pp65 epitope tested some years ago [30]. In this respect at least, some of these accumulated late-stage differentiated CD8$^+$ T cells may indeed be dysfunctional, potentially senescent, but this is not the case for all of them. Intriguingly, despite the loss of the expression of some costimulatory molecules and up-regulation of negative signaling receptors, it was demonstrated that polyfunctional CMV-specific CD8$^+$ T cells were at an intermediate state of differentiation and were not necessarily restricted in their replicative capacity by excessive telomere erosion [54]. As mentioned above, infection with CMV induces additionally the differentiation not only of CD4$^+$ T cells, but also NK cells - all of which share effector characteristics of CMV-specific CD8$^+$ T cells. It was supposed that this overlap in differentiation of these pools of lymphocytes might be regulated by shared transcriptional machinery [33]; thus, there could be a degree of interchangeability of immune elements maintaining control of CMV, and resulting in heterogeneous manifestations of the side-effects recognized as characteristic of immunosenescence.

It should not be forgotten that there is a second major type of T cells, those expressing γδ antigen receptors rather than the more common αβ receptors, which also display considerable age-related differences, mostly decreased in older adults. However, after infection with CMV, the usually numerically smaller pool of Vδ2-negative γδ T cells appears to be significantly increased and accumulates further throughout life (Fig. 4.2). In contrast, CMV-seropositivity has no influence on Vδ2-positive γδ T cells, which maintain a less differentiated phenotype [55]. Moreover, the γδ T-cell repertoire is more restricted in CMV-seropositive individuals. Taken together, these findings suggest that infection with CMV causes changes in γδ T-cell repertoire similar to those described in CD8$^+$ T cells [55, 56]. In a recent systems-level analysis of the immune system of healthy twins it was shown that CMV has indeed a major non-heritable impact on immune parameters (such as effector CD8$^+$ and γδ T cells) and can dramatically modulate the overall immune profile of healthy individuals [57].

It is conceivable that all these alterations might be important contributory factors to immunosenescence but certainly are not the whole story. Further functional and longitudinal studies are required to elucidate the relationship between ageing, CMV, immunosenescence and their clinical consequences [3, 41, 42]. However, a large body of mostly circumstantial and to a great extent controversial work has been published on the consequences of CMV infection for health, frailty and mortality.

4.4 Consequences for Health and Vaccination

People aged 60 and older represent over 11 % of the world population, with the proportion expected to increase to 22 % by 2050. Ageing is obviously associated with an increased frequency of age-related diseases including higher susceptibility to infections, cancer, cardiovascular and neurodegenerative diseases [58]. The role of CMV in the etiology of these age-associated diseases is currently under intensive investigation. It has been reported that CMV infection markedly increases mortality in the elderly and has been associated with frailty and impaired survival [59–61]. As discussed above, it is associated with the "Immune Risk Profile" in the Swedish OCTO/NONA studies but not in BELFRAIL. Functional decline of older individuals has been associated with immune parameters and the intensity of the response to CMV [62]. Thus, for example, in one powerful study, the impact of CMV infection on mortality was investigated in a cohort of 511 individuals aged at least 65 years at entry, who were then followed up for 18 years. Infection with CMV was associated with an increased mortality rate in healthy older individuals due to an excess of vascular deaths. It was estimated that those elderly who were CMV-seropositive at the beginning of the study had a near 4-year reduction in lifespan compared to those who were CMV-seronegative, a striking result with major implications for public health [59]. Other data, such as those from the large US NHANES-III survey, have shown that CMV seropositivity together with higher than median levels of the inflammatory marker CRP correlate with a significantly lower 10-year survival rate of individuals who were mostly middle-aged at the start of the study [63]. Further evidence comes from a recently published Newcastle 85+ study of the immune parameters of 751 octogenarians investigated for their power to predict survival during a 65-month follow-up. It was documented that CMV-seropositivity was associated with increased 6-year cardiovascular mortality or death from stroke and myocardial infarction. It was therefore concluded that CMV-seropositivity is linked to a higher incidence of coronary heart disease in octogenarians and that senescence in both the CD4+ and CD8+ T-cell compartments is a predictor of overall cardiovascular mortality [61]. Another study demonstrated that resting blood pressure is associated with the magnitude of CMV-specific CD8+ T-cell responses and with a novel regulatory type CD4+ T-cell subpopulation expressing CD25, CD39 and CD134 upon activation with CMV-antigen [64]. These investigators speculated that CMV infection might directly target vascular endothelium and smooth muscle and might be related to accelerated vascular pathology and mortality in the aged, and/or that CMV-specific T-cell immunity itself might directly contribute to this pathological process. The mechanism whereby CMV infection is associated with increased cardiovascular risk may relate to findings of increased arterial stiffness in patients with chronic kidney disease associated with CMV infection, which could contribute to the cardiovascular complications seen in these patients and possibly also reflect a more general CMV effect [65].

The issue concerning the role of CMV infection on mental health and cognitive state in the old remains even more controversial [14]. In a study of 1061 participants of the Lothian Birth Cohort a small but significant decrease in cognitive function was seen in the CMV-seropositive group. The likelihood of contracting CMV infection by age 70 is predicted by a number of demographic and environmental factors. After accounting for these, CMV infection (considered as serostatus) was not cognitively detrimental. Within CMV-seropositive individuals, however, higher CMV antibody levels were associated with lower general cognitive ability of old people [66].

In cancer, available evidence indicates that CMV is present in several solid tumors with a high prevalence, approaching 100 % in malignant glioblastoma [14]. The longer survival in patients whose tumors had low-grade CMV infection suggests that the level of CMV infection in malignant glioblastoma may be considered as a prognostic factor. Furthermore, it is possible that CMV may contribute to the pathogenesis of glioblastoma itself [67]. In a similar way, invasiveness and relapses of neuroblastoma may also be linked to the presence of human cytomegalovirus. The CMV-specific drug valganciclovir significantly reduced viral protein expression and tumor cell growth both in vitro and in vivo. Therefore, it was speculated that CMV infection may contribute to the pathogenesis of neuroblastoma and antiviral therapy may provide a novel treatment option for children with neuroblastoma [68]. Following these results, a randomized, double-blind, placebo-controlled, hypothesis-generating study was initiated to examine the safety and potential efficacy of valganciclovir as an add-on therapy for glioblastoma [69]. However, the involvement of immunity, if any, in these effects is not known.

CMV-seropositivity has been shown in some but not all studies to be linked to diminished humoral responsiveness of the old to influenza vaccination [70–74]. It was suggested that infection with CMV, accompanied by elevated pro-inflammatory potential, could contribute to the poor responsiveness to influenza vaccine in older individuals [73]. In two independent cohorts it was demonstrated that CD4+ T-cell responses to influenza core proteins are absent in almost half of CMV-seropositive older adults, whereas older people not infected with CMV respond as well as the young. It was concluded that advanced chronological age plays a role in compromised responses to influenza but only in concert with CMV infection [71]. Intriguing results from the recent study of Furman et al., indicated that CMV-seropositive young adults exhibited enhanced antibody responses to influenza vaccination, increased CD8+ T-cell sensitivity, and elevated levels of circulating interferon-gamma compared to seronegative individuals [45]. The overall decreased responses to vaccination in aged individuals commonly observed by others were also seen in this study, regardless of CMV status of the subjects. Results consistent with these findings were also reported in a murine model, where young mice infected with murine CMV also showed significant protection against influenza compared with uninfected mice. These data further show that CMV can have a beneficial effect on the immune response of young individuals—a finding that may go some way to explaining the ubiquity of CMV infection in human (and animal) populations [45]. These and other findings [43, 71] show that the CMV-associated impact on immune features phenotyped as immunosenescence is not necessarily detrimental to the host, and suggest that the term "senescence" should be avoided because this by definition means something detrimental. Rather, data suggest that the remodeling of the T-cell compartment in the presence of a latent infection with CMV represents a crucial adaptation of the immune system towards the chronic challenge of lifelong CMV. Thus, continued immunosurveillance against persistent CMV appears to be more important for host survival than reserving immune resources for responses to other viruses [71]. These data add to the accumulating evidence that infection with CMV has important but apparently very heterogeneous effects on responses to other viruses and this may influence on the design of influenza (and other) vaccines, especially for the elderly population.

Acknowledgments This work was supported by the European Commission under Grant Agreement FP7 259679, Integrated research on developmental determinants of ageing and longevity, "IDEAL" and by an unrestricted educational grant from the Croeni Foundation (to GP).

References

1. Müller L, Pawelec G. As we age: does slippage of quality control in the immune system lead to collateral damage? Ageing Res Rev. 2015;23(Pt A):116–23.
2. Pawelec G. Immunosenescence: role of cytomegalovirus. Exp Gerontol. 2014;54:1–5.
3. Fülöp T, Larbi A, Pawelec G. Human T cell aging and the impact of persistent viral infections. Front Immunol. 2013;4:271.
4. Pawelec G. Hallmarks of human "immunosenescence": adaptation or dysregulation? Immunol Ageing. 2012;9(1):15.
5. Arens R, Remmerswaal EB, Bosch JA, van Lier RA. 5(th) International Workshop on CMV and Immunosenescence—a shadow of cytomegalovirus infection on immunological memory. Eur J Immunol. 2015;45(4):954–7.
6. Meijers RW, Litjens NH, Hesselink DA, Langerak AW, Baan CC, Betjes MG. Primary cytomegalovirus infection significantly impacts circulating T cells in kidney transplant recipients. Am J Transplant. 2015;15(12):3143–56.
7. Miles DJ, van der Sande M, Jeffries D, Kaye S, Ismaili J, Ojuola O, et al. Cytomegalovirus infection in Gambian infants leads to profound CD8 T-cell differentiation. J Virol. 2007;81(11):5766–76.
8. Dowd JB, Aiello AE, Alley DE. Socioeconomic disparities in the seroprevalence of cytomegalovirus infection in the US population: NHANES III. Epidemiol Infect. 2009;137(1):58–65.
9. Sijmons S, Van Ranst M, Maes P. Genomic and functional characteristics of human cytomegalovirus revealed by next-generation sequencing. Viruses. 2014;6(3):1049–72.
10. Boeckh M, Geballe AP. Cytomegalovirus: pathogen, paradigm, and puzzle. J Clin Invest. 2011;121(5):1673–80.
11. Hanley PJ, Bollard CM. Controlling cytomegalovirus: helping the immune system take the lead. Viruses. 2014;6(6):2242–58.
12. Britt W. Manifestations of human cytomegalovirus infection: proposed mechanisms of acute and chronic disease. Curr Top Microbiol Immunol. 2008;325:417–70.

13. Effros RB. The silent war of CMV in aging and HIV infection. Mech Ageing Dev. 2015;pii:S0047-6374(15)30014-2.

14. Sansoni P, Vescovini R, Fagnoni FF, Akbar A, Arens R, Chiu YL, et al. New advances in CMV and immunosenescence. Exp Gerontol. 2014;55:54–62.

15. Kalejta RF. Tegument proteins of human cytomegalovirus. Microbiol Mol Biol Rev. 2008;72(2):249–65. table of contents.

16. Nachmani D, Lankry D, Wolf DG, Mandelboim O. The human cytomegalovirus microRNA miR-UL112 acts synergistically with a cellular microRNA to escape immune elimination. Nat Immunol. 2010;11(9):806–13.

17. Wilkinson GW, Tomasec P, Stanton RJ, Armstrong M, Prod'homme V, Aicheler R, et al. Modulation of natural killer cells by human cytomegalovirus. J Clin Virol. 2008;41(3):206–12.

18. Slobedman B, Barry PA, Spencer JV, Avdic S, Abendroth A. Virus-encoded homologs of cellular interleukin-10 and their control of host immune function. J Virol. 2009;83(19):9618–29.

19. Beisser PS, Lavreysen H, Bruggeman CA, Vink C. Chemokines and chemokine receptors encoded by cytomegaloviruses. Curr Top Microbiol Immunol. 2008;325:221–42.

20. McCormick AL. Control of apoptosis by human cytomegalovirus. Curr Top Microbiol Immunol. 2008;325:281–95.

21. Reeves M, Sinclair J. Aspects of human cytomegalovirus latency and reactivation. Curr Top Microbiol Immunol. 2008;325:297–313.

22. Reeves M, Sinclair J. Regulation of human cytomegalovirus transcription in latency: beyond the major immediate-early promoter. Viruses. 2013;5(6):1395–413.

23. Lee SH, Albright ER, Lee JH, Jacobs D, Kalejta RF. Cellular defense against latent colonization foiled by human cytomegalovirus UL138 protein. Sci Adv. 2015;1(10), e1501164.

24. Poole E, Sinclair J. Sleepless latency of human cytomegalovirus. Med Microbiol Immunol. 2015;204(3):421–9.

25. Liu XF, Wang X, Yan S, Zhang Z, Abecassis M, Hummel M. Epigenetic control of cytomegalovirus latency and reactivation. Viruses. 2013;5(5):1325–45.

26. Goodrum F, Caviness K, Zagallo P. Human cytomegalovirus persistence. Cell Microbiol. 2012;14(5):644–55.

27. Derhovanessian E, Maier AB, Beck R, Jahn G, Hahnel K, Slagboom PE, et al. Hallmark features of immunosenescence are absent in familial longevity. J Immunol. 2010;185(8):4618–24.

28. Karrer U, Sierro S, Wagner M, Oxenius A, Hengel H, Koszinowski UH, et al. Memory inflation: continuous accumulation of antiviral CD8+ T cells over time. J Immunol. 2003;170(4):2022–9.

29. Kim J, Kim AR, Shin EC. Cytomegalovirus Infection and Memory T Cell Inflation. Immune Netw. 2015;15(4):186–90.

30. Ouyang Q, Wagner WM, Voehringer D, Wikby A, Klatt T, Walter S, et al. Age-associated accumulation of CMV-specific CD8+ T cells expressing the inhibitory killer cell lectin-like receptor G1 (KLRG1). Exp Gerontol. 2003;38(8):911–20.

31. Jackson SE, Mason GM, Okecha G, Sissons JG, Wills MR. Diverse specificities, phenotypes, and antiviral activities of cytomegalovirus-specific CD8+ T cells. J Virol. 2014;88(18):10894–908.

32. Gibson L, Dooley S, Trzmielina S, Somasundaran M, Fisher D, Revello MG, et al. Cytomegalovirus (CMV) IE1- and pp 65-specific CD8+ T cell responses broaden over time after primary CMV infection in infants. J Infect Dis. 2007;195(12):1789–98.

33. Vieira Braga FA, Hertoghs KM, van Lier RA, van Gisbergen KP. Molecular characterization of HCMV-specific immune responses: Parallels between CD8(+) T cells, CD4(+) T cells, and NK cells. Eur J Immunol. 2015;45(9):2433–45.

34. Stowe RP, Kozlova EV, Yetman DL, Walling DM, Goodwin JS, Glaser R. Chronic herpesvirus reactivation occurs in aging. Exp Gerontol. 2007;42(6):563–70.

35. Beswick M, Pachnio A, Lauder SN, Sweet C, Moss PA. Antiviral therapy can reverse the development of immune senescence in elderly mice with latent cytomegalovirus infection. J Virol. 2013;87(2):779–89.

36. Mekker A, Tchang VS, Haeberli L, Oxenius A, Trkola A, Karrer U. Immune senescence: relative contributions of age and cytomegalovirus infection. PLoS Pathog. 2012;8(8), e1002850.

37. Pachnio A, Begum J, Fox A, Moss P. Acyclovir therapy reduces the CD4+ T cell response against the immunodominant pp 65 protein from cytomegalovirus in immune competent individuals. PLoS One. 2015;10(4), e0125287.
38. Parry HM, Zuo J, Frumento G, Mirajkar N, Inman C, Edwards E, et al. Cytomegalovirus viral load within blood increases markedly in healthy people over the age of 70 years. Immunol Ageing. 2016;13:1.
39. Sylwester AW, Mitchell BL, Edgar JB, Taormina C, Pelte C, Ruchti F, et al. Broadly targeted human cytomegalovirus-specific CD4+ and CD8+ T cells dominate the memory compartments of exposed subjects. J Exp Med. 2005;202(5):673–85.
40. Pawelec G, Derhovanessian E, Larbi A, Strindhall J, Wikby A. Cytomegalovirus and human immunosenescence. Rev Med Virol. 2009;19(1):47–56.
41. Pawelec G, McElhaney JE, Aiello AE, Derhovanessian E. The impact of CMV infection on survival in older humans. Curr Opin Immunol. 2012;24(4):507–11.
42. Pawelec G, Derhovanessian E. Role of CMV in immune senescence. Virus Res. 2011; 157(2):175–9.
43. Adriaensen W, Derhovanessian E, Vaes B, Van Pottelbergh G, Degryse JM, Pawelec G, et al. CD4:8 ratio >5 is associated with a dominant naive T-cell phenotype and impaired physical functioning in CMV-seropositive very elderly people: results from the BELFRAIL study. J Gerontol A Biol Sci Med Sci. 2015;70(2):143–54.
44. Pera A, Campos C, Corona A, Sanchez-Correa B, Tarazona R, Larbi A, et al. CMV latent infection improves CD8+ T response to SEB due to expansion of polyfunctional CD57+ cells in young individuals. PLoS One. 2014;9(2), e88538.
45. Furman D, Jojic V, Sharma S, Shen-Orr SS, Angel CJ, Onengut-Gumuscu S, et al. Cytomegalovirus infection enhances the immune response to influenza. Sci Transl Med. 2015;7(281):281ra43.
46. Barton ES, White DW, Cathelyn JS, Brett-McClellan KA, Engle M, Diamond MS, et al. Herpesvirus latency confers symbiotic protection from bacterial infection. Nature. 2007;447(7142):326–9.
47. Wertheimer AM, Bennett MS, Park B, Uhrlaub JL, Martinez C, Pulko V, et al. Aging and cytomegalovirus infection differentially and jointly affect distinct circulating T cell subsets in humans. J Immunol. 2014;192(5):2143–55.
48. Turner JE, Campbell JP, Edwards KM, Howarth LJ, Pawelec G, Aldred S, et al. Rudimentary signs of immunosenescence in Cytomegalovirus-seropositive healthy young adults. Age (Dordr). 2014;36(1):287–97.
49. Gamadia LE, van Leeuwen EM, Remmerswaal EB, Yong SL, Surachno S, Wertheim-van Dillen PM, et al. The size and phenotype of virus-specific T cell populations is determined by repetitive antigenic stimulation and environmental cytokines. J Immunol. 2004;172(10):6107–14.
50. Redeker A, Welten SP, Arens R. Viral inoculum dose impacts memory T-cell inflation. Eur J Immunol. 2014;44(4):1046–57.
51. O'Hara GA, Welten SP, Klenerman P, Arens R. Memory T cell inflation: understanding cause and effect. Trends Immunol. 2012;33(2):84–90.
52. Terrazzini N, Bajwa M, Vita S, Thomas D, Smith H, Vescovini R, et al. Cytomegalovirus infection modulates the phenotype and functional profile of the T-cell immune response to mycobacterial antigens in older life. Exp Gerontol. 2014;54:94–100.
53. McElhaney JE, Zhou X, Talbot HK, Soethout E, Bleackley RC, Granville DJ, et al. The unmet need in the elderly: how immunosenescence, CMV infection, co-morbidities and frailty are a challenge for the development of more effective influenza vaccines. Vaccine. 2012;30(12):2060–7.
54. Riddell NE, Griffiths SJ, Rivino L, King DC, Teo GH, Henson SM, et al. Multifunctional cytomegalovirus (CMV)-specific CD8(+) T cells are not restricted by telomere-related senescence in young or old adults. Immunology. 2015;144(4):549–60.
55. Roux A, Mourin G, Larsen M, Fastenackels S, Urrutia A, Gorochov G, et al. Differential impact of age and cytomegalovirus infection on the gammadelta T cell compartment. J Immunol. 2013;191(3):1300–6.

56. Alejenef A, Pachnio A, Halawi M, Christmas SE, Moss PA, Khan N. Cytomegalovirus drives Vdelta2neg gammadelta T cell inflation in many healthy virus carriers with increasing age. Clin Exp Immunol. 2014;176(3):418–28.
57. Brodin P, Jojic V, Gao T, Bhattacharya S, Angel CJ, Furman D, et al. Variation in the human immune system is largely driven by non-heritable influences. Cell. 2015;160(1-2):37–47.
58. Pera A, Campos C, Lopez N, Hassouneh F, Alonso C, Tarazona R, et al. Immunosenescence: Implications for response to infection and vaccination in older people. Maturitas. 2015;82(1):50–5.
59. Savva GM, Pachnio A, Kaul B, Morgan K, Huppert FA, Brayne C, et al. Cytomegalovirus infection is associated with increased mortality in the older population. Aging Cell. 2013;12(3):381–7.
60. Wikby A, Ferguson F, Forsey R, Thompson J, Strindhall J, Lofgren S, et al. An immune risk phenotype, cognitive impairment, and survival in very late life: impact of allostatic load in Swedish octogenarian and nonagenarian humans. J Gerontol A Biol Sci Med Sci. 2005;60(5):556–65.
61. Spyridopoulos I, Martin-Ruiz C, Hilkens C, Yadegarfar ME, Isaacs J, Jagger C, et al. CMV seropositivity and T-cell senescence predict increased cardiovascular mortality in octogenarians: results from the Newcastle 85+ study. Aging Cell. 2015.
62. Moro-Garcia MA, Alonso-Arias R, Lopez-Vazquez A, Suarez-Garcia FM, Solano-Jaurrieta JJ, Baltar J, et al. Relationship between functional ability in older people, immune system status, and intensity of response to CMV. Age (Dordr). 2012;34(2):479–95.
63. Simanek AM, Dowd JB, Pawelec G, Melzer D, Dutta A, Aiello AE. Seropositivity to cytomegalovirus, inflammation, all-cause and cardiovascular disease-related mortality in the United States. PLoS One. 2011;6(2), e16103.
64. Terrazzini N, Bajwa M, Vita S, Cheek E, Thomas D, Seddiki N, et al. A novel cytomegalovirus-induced regulatory-type T-cell subset increases in size during older life and links virus-specific immunity to vascular pathology. J Infect Dis. 2014;209(9):1382–92.
65. Wall NA, Chue CD, Edwards NC, Pankhurst T, Harper L, Steeds RP, et al. Cytomegalovirus seropositivity is associated with increased arterial stiffness in patients with chronic kidney disease. PLoS One. 2013;8(2), e55686.
66. Gow AJ, Firth CM, Harrison R, Starr JM, Moss P, Deary IJ. Cytomegalovirus infection and cognitive abilities in old age. Neurobiol Aging. 2013;34(7):1846–52.
67. Rahbar A, Orrego A, Peredo I, Dzabic M, Wolmer-Solberg N, Straat K, et al. Human cytomegalovirus infection levels in glioblastoma multiforme are of prognostic value for survival. J Clin Virol. 2013;57(1):36–42.
68. Wolmer-Solberg N, Baryawno N, Rahbar A, Fuchs D, Odeberg J, Taher C, et al. Frequent detection of human cytomegalovirus in neuroblastoma: a novel therapeutic target? Int J Cancer. 2013;133(10):2351–61.
69. Stragliotto G, Rahbar A, Solberg NW, Lilja A, Taher C, Orrego A, et al. Effects of valganciclovir as an add-on therapy in patients with cytomegalovirus-positive glioblastoma: a randomized, double-blind, hypothesis-generating study. Int J Cancer. 2013;133(5):1204–13.
70. Frasca D, Blomberg BB. Aging, cytomegalovirus (CMV) and influenza vaccine responses. Hum Vaccin Immunother. 2015;12(3):682–90.
71. Derhovanessian E, Maier AB, Hahnel K, McElhaney JE, Slagboom EP, Pawelec G. Latent infection with cytomegalovirus is associated with poor memory CD4 responses to influenza A core proteins in the elderly. J Immunol. 2014;193(7):3624–31.
72. Derhovanessian E, Theeten H, Hahnel K, Van Damme P, Cools N, Pawelec G. Cytomegalovirus-associated accumulation of late-differentiated CD4 T-cells correlates with poor humoral response to influenza vaccination. Vaccine. 2013;31(4):685–90.
73. Trzonkowski P, Mysliwska J, Szmit E, Wieckiewicz J, Lukaszuk K, Brydak LB, et al. Association between cytomegalovirus infection, enhanced proinflammatory response and low level of anti-hemagglutinins during the anti-influenza vaccination—an impact of immunosenescence. Vaccine. 2003;21(25-26):3826–36.
74. Kim TS, Sun J, Braciale TJ. T cell responses during influenza infection: getting and keeping control. Trends Immunol. 2011;32(5):225–31.

Effects of Ageing on the Vaccination Response

5

Birgit Weinberger

Abstract

Vaccination is the most effective measure to prevent infectious diseases, as vaccination of children has proven world-wide. The topic of vaccines for the older adult is receiving increased attention due to demographic changes and the increased incidence and severity of many infections in older adults. Vaccine recommendations for the old are implemented in most countries and include vaccination against influenza, *Streptococcus pneumoniae* and frequently also herpes zoster. Vaccines which are recommended for all adults, e.g. against tetanus, diphtheria and pertussis, need to be considered for the old. However, vaccination coverage is frequently poor. Immunogenicity and clinical efficacy of most vaccines decrease with age and are therefore not optimal in the old. Strategies to improve vaccines for the older age group include; optimized schedules, high-dose formulations, alternative routes of administration, such as intradermal vaccination, and the use of adjuvants. Detailed knowledge of age-associated changes to the immune system will enable us to rationally design new vaccines, which specifically target the aged immune system. The development of novel vaccines replacing existing formulations and of vaccines against additional pathogens is being actively pursued, and together with optimal use of existing vaccines, will contribute to improved health and quality of life for an ageing society.

Keywords

Vaccination • Ageing • Influenza • *Streptococcus pneumoniae* • Herpes zoster • Tetanus • Diphtheria • Pertussis • Respiratory syncytial virus

B. Weinberger (✉)
Research Institute for Biomedical Aging Research, University of Innsbruck,
Rennweg 10, 6020 Innsbruck, Austria
e-mail: birgit.weinberger@uibk.ac.at

© Springer International Publishing Switzerland 2017
V. Bueno et al. (eds.), *The Ageing Immune System and Health*,
DOI 10.1007/978-3-319-43365-3_5

5.1   Introduction

The incidence and severity of many infections are increased in older adults. Influenza causes approximately 36,000 deaths and more than 100,000 hospitalizations in the USA every year, the majority of which occur in persons older than 65 years [1, 2]. The incidence of herpes zoster is particularly high in the old and it has been estimated that up to 50% of all cases affect people >85 years of age [3]. Pneumonia and invasive disease (bacteremia, meningitis, etc.) caused by *Streptococcus pneumoniae*, group B streptococci (GBS) and other bacteria are also prevalent in the older population [4–6]. In addition, urinary tract infections as well as infections of the skin and soft tissue frequently affect the old [7, 8]. The reasons for this increased risk include underlying chronic disease, anatomical changes e.g. in the skin, lung and urinary tract, medical procedures, such as surgery and catheters, as well as altered immune function in old age [9]. The most effective measure to prevent infectious disease is vaccination. The outstanding success of childhood vaccination is undoubted, but the importance of vaccination beyond the paediatric age is frequently underestimated. Over the last 20–30 years tremendous progress has been achieved in developing novel/improved vaccines for children, but a lot of work still needs to be done to optimize vaccines for the elderly.

5.2   Vaccines Specifically Recommended for Older Adults

Vaccination recommendations for adults and the elderly are established in many countries. Vaccination against influenza and *S. pneumoniae* are frequently recommended for older adults and persons with underlying diseases, and many countries also advocate vaccination against herpes zoster (Table 5.1). The age limits for these recommendations vary, e.g. between ≥50 years and ≥65 years in Europe for influenza vaccination [10]. Regular booster immunizations against tetanus/diphtheria and in some cases pertussis are recommended for adults including the elderly in many countries. Vaccine uptake differs tremendously between European countries with more than 70% of the older population being vaccinated against influenza in The Netherlands and the United Kingdom, but below 10% in Poland, Latvia and Estonia during the 2012–2013 season [10]. Data for vaccination coverage are frequently difficult to obtain for adults, particularly as not all countries/health systems offer financial coverage for adult vaccination.

5.2.1   Influenza Vaccine

Most currently used influenza vaccines are composite vaccines containing disrupted inactivated viral particles or subunit vaccines, for which the viral surface proteins hemagglutinin (HA) and neuraminidase are further enriched by purification steps. These vaccines are routinely produced in embryonated chicken eggs. In addition, vaccines produced in cell culture and a recombinant vaccine containing HA proteins expressed in insect cells using a baculovirus expression system are available. Over

Table 5.1 Vaccination recommendations for adults and older adults in selected countries for 2015

	USA[a]	Germany[b]	Austria[c]	UK[d]	Italy[e]
Influenza	Annually for all adults	Annually over 60	Annually for all adults, particularly over 50	Annually over 65	Annually for all adults, particularly over 65
S. pneumoniae[f]	Once over 50 PCV13, after 1 year PPV23	Once over 60 PPV23	Once over 50 PCV13, after 1 year PPV23	Once over 65 PPV23	Once over 65 PCV13, followed by PPV23
Herpes zoster	Once over 60	–	Once over 50	Once over 70	Once over 60
Diphtheria[g]	Every 10 years	Every 10 years	Every 10 years, over 60 every 5 years	–	Every 10 years
Tetanus[g]	Every 10 years	Every 10 years	Every 10 years, over 60 every 5 years	–	Every 10 years
Pertussis (acellular)[g]	Once during adulthood	Once during adulthood	Every 10 years, over 60 every 5 years	–	Every 10 years
Polio (inactivated)[g]	–	–	Every 10 years, over 60 every 5 years	–	

General recommendations for age groups are shown, additional recommendations for specific risk groups, e.g. persons with underlying diseases are not included here

[a]http://www.cdc.gov/vaccines/schedules/downloads/adult/adult-schedule.pdf

[b]http://www.rki.de/DE/Content/Infekt/EpidBull/Archiv/2015/Ausgaben/34_15.pdf?blob=publicationFile

[c]http://bmg.gv.at/cms/home/attachments/8/9/4/CH1100/CMS1389365860013/impfplan.pdf

[d]https://www.gov.uk/government/uploads/system/uploads/attachment_data/file/473570/9406_PHE_2015_Complete_Immunisation_Schedule_A4_21.pdf

[e]http://www.quotidianosanita.it/allegati/allegato1955037.pdf

[f]For persons without prior vaccination with PCV13 or PPV23. Detailed recommendations for persons who have previously received either vaccine are given in the cited documents

[g]For persons with adequate primary vaccination earlier in life

the last few decades influenza vaccines have contained three different strains (A/H1N1, A/H3N2, B). The exact composition of the vaccine is determined each year by the World Health Organization based on surveillance data and predictions on circulating strains. For many years, two B strains have been circulating in parallel, leading to a frequent mismatch between the vaccine and the predominant circulating strain [11]. Recently, quadrivalent vaccines containing two B strains have been developed and licensed by several vaccine manufacturers [12, 13].

Influenza vaccines have been extensively studied in the aged and a plethora of studies have shown that immunogenicity—measured by hemagglutination inhibition assay (HAI)—of subunit and split influenza vaccines is lower in old compared to young adults [14, 15]. In addition to age, health status can also influence immune responses to influenza, as HAI titers are significantly higher in the healthy old compared to those with chronic diseases [16]. Most studies additionally investigate rates of seroprotection (HAI $\geq$1:40) and seroconversion (titer increase $\geq$4-fold), which are frequently also lower in the elderly. A meta-analysis of 31 studies demonstrated unadjusted odds ratios (OR) of 0.48 (95 % CI 0.41–0.55; H1N1), 0.63 (0.55–0.73; H3N2), and 0.38 (0.33–0.44; B) for seroconversion and 0.47 (0.40–0.55; H1N1), 0.53 (0.45–0.63; H2N3), and 0.58 (0.50–0.67; B) for seroprotection in a comparison of old versus young adults [17]. However, HAI titers are imperfect correlates of protection, as laboratory-confirmed influenza infections have been documented despite the presence of HAI titers $\geq$1:640 [18]. Age-related changes in the antibody repertoire after influenza vaccination have been reported [19, 20], but other studies showed intact quality of influenza-specific antibodies in older adults [21]. Similar to the humoral immune response, cell-mediated immunity after vaccination is lower in old compared to young adults [22, 23].

Studies analyzing a formulation's clinical efficacy or effectiveness against influenza are difficult to compare, as their outcome heavily depends on various parameters. Studies enrolling persons over the age of 65 can substantially differ regarding the age distribution and the median age. The health status and underlying co-morbidities of the participants are also of importance. In older patients, frailty, a multifactorial syndrome characterized by a reduced stress resistance and physiological reserve [24], impacts susceptibility to influenza and responsiveness to influenza vaccine [25]. These parameters can be allowed for by inclusion/exclusion criteria and are definitely discordant in community-dwelling versus institutionalized cohorts. The living situation also has an impact on transmission patterns and prevalence of influenza, which are different in nursing homes compared to the general community. Other epidemiological parameters, e.g. virulence of the virus, prevalence in the population and the degree of mismatch between the vaccine strains and circulating virus strains [26] differ from year to year, thus making meta-analyses more complicated. Another critical parameter is the clinical read-out. A commonly used read-out parameter is influenza-like illness (ILI). The Center for Disease Control and Prevention (CDC) provides a definition of ILI as fever with either cough or sore throat [27] which has a high positive predictive value for influenza infection in young adults (86.8 %) during influenza season [28]. Influenza symptoms in older adults are frequently atypical and not associated with high fever. Consequently, many influenza infections might be missed by this clinical assessment. On the other hand, other pathogens can cause symptoms similar to influenza and the resulting infections might be miss-classified as influenza infection. One study on a hospitalized cohort (median age 60 year, range 15–99 year) reported ILI symptoms for only half of the patients with laboratory-confirmed influenza [29]. Laboratory confirmation of influenza infection can be achieved by detection of the virus in culture or by polymerase chain reaction (PCR). Both methods have limited sensitivity [30, 31] leading to an underestimation of influenza cases. Randomized, placebo-controlled trials are rare but two studies showed significant efficacy of influenza vaccination [32, 33]. Another study did not confirm these results [34]. Placebo-controlled trials are no longer considered ethical, as influenza vaccination is recommended for the elderly. Several systematic reviews and meta-analyses have estimated the clinical efficacy and/or effectiveness of a given influenza vaccine, taking into consideration not only randomized trials, but also cohort and case-control studies. It can be concluded that protection is lower in the old than in young adults [35–37].

The standard trivalent inactivated vaccines (SD-TIV) containing 15 µg HA per strain are obviously not optimal for the older population. Several approaches have been pursued in order to improve influenza vaccines for the elderly. Two-dose strategies using SD-TIV were not successful. Administration of two doses of SD-TIV 84 days apart to older residents of long-term care facilities did not increase geometric mean titers (GMT) nor seroprotection rates [38]. This study, however, reported improved immune responses to a high-dose TIV (30 µg HA per strain) and a beneficial effect of a standard dose booster shot for participants who received the high-dose primary vaccination. The concept of increased antigen dosage has been studied in more detail and it has been demonstrated that higher antigen doses lead to higher antibody titers and seroprotection rates [39, 40]. A trivalent inactivated vaccine containing 60 µg HA per strain has been licensed for persons older than 65 years in the US. Retrospective cohort studies reported controversial data regarding clinical efficacy of the high-dose vaccine compared to SD-TIV [41, 42], but a randomized trial enrolling more than 31,000 old participants demonstrated that the high-dose vaccine was 24.2% more efficacious against laboratory-confirmed ILI compared to SD-TIV [43].

Alternative routes of administration have also been considered in order to improve vaccine efficacy. As the influenza virus enters the body via the airways, mucosal immunity plays an important role in the defense against influenza. Vaccination at the site of entry seems therefore likely to be a promising strategy. A live-attenuated vaccine which is administered intra-nasally is licensed for persons aged 2–49 years, but not for older age groups. As the skin harbors many innate immune cells including dendritic cells (DC), monocytes, macrophages and accessory cells such as keratinocytes [44], it is an attractive site for vaccine delivery. Intradermally administered vaccine antigens can be taken up by DC residing in the dermis or epidermis, which transport the antigen to the draining lymph nodes [45]. Intradermal delivery of standard dose (15 µg HA) influenza vaccine elicits higher antibody titers and seroprotection rates in older persons compared to the intramuscular route [46]. This vaccine uses a specialized ready-to-use microinjection device, which ensures correct application into the dermis.

Aluminum salts have been used as adjuvants with many vaccines for several decades and have been the only adjuvants available for human vaccines for a long time. However, they show limited effects when used in combination with inactivated influenza vaccines. MF59, an oil-in-water based adjuvant containing squalene, was the first novel adjuvant to be developed and licensed in combination with influenza antigens [47]. Its mode of action has been extensively studied and it has been concluded that MF59 modifies the local environment at the injection site inducing chemokine production, which leads to the recruitment of innate immune cells and as a consequence efficient uptake of the antigen and transport to the lymph nodes [48]. The immunogenicity of MF59-adjuvanted vaccine compared to conventional TIV has been analyzed in many studies, and meta-analyses confirmed slightly but still significantly increased geometric mean titers measured by HAI for older adults, in particular for elderly with underlying diseases [49, 50]. The clinical efficacy of MF59-adjuvanted vaccine in comparison to standard TIV was analyzed in a large study in Italy. The risk of hospitalization for influenza or pneumonia was 25 % lower for the adjuvanted vaccine (relative risk 0.75, 95 % CI 0.57–0.98) [51]. In a study including residents of long-term care facilities the risk of ILI was higher in persons receiving standard TIV (OR 1.52, 95 % CI 1.22–1.88) than in those who had received the adjuvanted vaccine. This effect was even more pronounced in patients with respiratory and cardiovascular disease [52]. In seasons with a high degree of mismatch between the vaccine strain and the circulating viral strain vaccine efficacy can drop below 30 % [53]. Several clinical studies have demonstrated that MF59-adjuvanted vaccine induces substantially higher antibody responses against heterologous viral strains compared to standard TIV [54–56]. This broader neutralizing activity, together with influenza-specific CD4+ T cells which are elicited by MF59-containing influenza vaccine [57], might contribute substantially to the improved clinical efficacy observed with the adjuvanted vaccine. Several studies showed that antibodies elicited by the adjuvanted vaccine recognize different epitopes of the viral HA and NA proteins. The underlying mechanisms of this phenomenon are still under investigation [58].

Other oil-in-water based adjuvants which have been licensed for influenza vaccines are AF03, which is also based on squalene, for the pandemic H1N1 vaccine [59] and AS03 additionally containing α-tocopherol for the pandemic A/H1N1 and the avian A/H5N1 vaccine [60]. Other adjuvants such as Toll-like receptor agonists, virus-like particles, liposomes etc. are at different stages of development in combination with influenza antigens [58]. The only other licensed adjuvanted vaccine against influenza uses virosomes, which are reconstituted viral envelopes containing viral HA and NA proteins, but lack viral RNA. This adjuvanted influenza vaccine has been shown to elicit slightly higher antibody titers compared to conventional TIV [61].

For many years the idea of a "universal" influenza vaccine, which would be able to induce long-lasting immunity, protect from all strains of influenza, and thereby solve the issue of annual re-vaccination, has inspired numerous researchers. The first step in its development is the selection of antigens. Conserved regions of the surface proteins HA and NA as well as internal viral proteins such as the matrix protein M2 and the nuclear protein NP have been suggested as potential candidate antigens. The second step is the selection of a vaccine platform, which determines the type of immune response (antibodies, CD4+ or CD8+ T cells) and possible routes of administration. Viral vectors, DNA vaccines, virus-like particles and also the use of existing or novel adjuvants are discussed in this context (summarized in [62]).

5.2.2 Pneumococcal Vaccine

Infections with *Streptococcus pneumoniae* are a significant cause for morbidity and mortality in the old and account for 25–35 % of bacterial pneumonias requiring hospitalization [63]. The 23-valent pneumococcal polysaccharide vaccine (PPV23) has been used for many years in the older population. As PPV23 contains plain capsular polysaccharides, it elicits only T cell-independent antibody responses resulting in a lack of immunological memory and booster effects upon repeated vaccination. The efficacy of PPV23 to prevent pneumococcal disease is controversially discussed. A meta-analysis reported efficacy against invasive disease (OR 0.26; 95 % CI 0.15–0.46), and to a lesser extent all-cause pneumonia (OR 0.71; 95 % CI 0.52–0.97) in adults [64]. However, the clinical efficacy of PPV23 against pneumonia in older persons is frequently doubted. Some individual studies reported efficacy against pneumonia in institutionalized elderly or healthy elderly, respectively [65, 66], but a meta-analysis did not confirm these findings and showed only non-significant vaccine efficacy of 16 % (95 % CI-50–53) [67]. Heterogeneous study populations, such as institutionalized versus community-dwelling elderly, as well as varying read-out parameters (pneumococcal pneumonia, all-cause pneumonia, hospitalization due to pneumonia etc), complicate meta-analyses of pneumococcal vaccine efficacy (reviewed in [68]). Serotype-specific antibody concentrations, which are accepted as correlates of protection, are lower in the elderly and in individuals with underlying disease [63, 69]. Protein-conjugated vaccines (PCV) have been developed for childhood vaccination, as PPV-23 does not elicit immune responses in infants. A 7-valent conjugated vaccine (PCV7) has been successfully used in children and is capable of eliciting memory responses. PCV7 induces higher antibody concentrations in the old compared to PPV23 and shows a booster effect following a second dose of vaccine after 1 year [70]. Routine vaccination of children with PCV7 leads to a decrease of transmission of *S. pneumoniae* and thereby provides indirect protection of the old against the serotypes included in the vaccine [71]. In the following years conjugated vaccines covering more pneumococcal serotypes have been developed for the pediatric market (PCV10 and PCV13) and replaced PCV7 for childhood vaccination. PCV13 has also been licensed for adults and recent vaccination recommendations for the old include a single vaccination with PCV13 in many countries, with an additional dose of PCV23 12 months later in some countries. Immunogenicity of PCV13 has been demonstrated in adults who had never received PPV23 [72] as well as in individuals who had received PPV23 several years earlier. Clinical efficacy of PCV13 in older adults has been demonstrated in a large randomized, double-blind, placebo-controlled study. More than 84,000 pneumococcal vaccine-naïve persons above 65 years of age received one dose of PCV13 or placebo, and were followed up for several years for community-acquired pneumonia (CAP) and invasive pneumococcal disease (IPD). Detection of *S. pneumoniae* and identification of vaccine serotypes was performed by culture and/or urinary antigen test. Vaccine efficacy was 45.6 % (95.2 % CI 21.8–62.5, $p < 0.001$) for confirmed vaccine-type CAP. Efficacy was also demonstrated for vaccine-type non-bacteraemic and noninvasive CAP (45.0 %, 95.2 % CI 14.2–65.3, $p < 0.01$) and vaccine-type IPD (75.0 %, 95 % CI 41.4–90.8, $p < 0.001$) [73]. In a post-hoc analysis of this study the effect of age on vaccine efficacy was studied and the statistical model showed a decline of vaccine efficacy for vaccine-type CAP and IPD from 65 % (95 % CI 38–81) in 65-year old subjects, to 40 % (95 % CI 17–56) in 75-year old subjects [74]. Recently, safety and immunogenicity of a 15-valent conjugate vaccine, which includes two additional serotypes, has been demonstrated in healthy adults [75].

Polysaccharide and polysaccharide-conjugate vaccines share the limitation of restricted serotype coverage. After the introduction of PCV7, serotype replacement has been observed, i.e. the number of disease cases caused by pneumococcal serotypes not included in the vaccine increased [76]. Similar effects start to emerge for PCV13 [77]. Several approaches aim to develop a universal vaccine, which elicits serotype-independent immune responses and thereby protects from all pneumococcal strains. Several pneumococcal proteins have been identified as vaccine candidates, as they are highly conserved and expressed by all clinical isolates and elicit protective, long-lasting cellular and humoral immunity in animal models. Alternative approaches include combinations of PCV with pneumococcal proteins, whole cell inactivated vaccines and live-attenuated vaccines (summarized in [68]).

5.2.3   Herpes Zoster Vaccine

Almost 100 % of the adult population are latently infected with varicella zoster virus (VZV), which manifests as chickenpox during the primary infection usually occurring in childhood, and as herpes zoster (shingles) during reactivation. It is believed that reactivation occurs when VZV-specific cellular immune responses decline due to immunosuppressive conditions or immunosenescence, and less VZV-specific T cells are detectable in the old compared to young adults [78]. The incidence of herpes zoster increases with age and rises from 1 to 3 per 1000 person-years in young adults to 6–10 per 1000 person-years in persons aged 60–70 years. The incidence is even higher in the very old [3]. The lifetime risk of a herpes zoster infection is approximately 50 % among those who reach 85 years of age [79]. Post-herpetic neuralgia (PHN) is characterized by persistent pain for months or even years after acute herpes zoster and is the most common complication of herpes zoster. PHN can have a dramatic impact on activities of daily living, frequently leading to loss of independence and institutionalization in the old [80]. A vaccine containing a high dose of the live-attenuated Oka-strain, the strain which is also used as a paediatric vaccine against chickenpox, was licensed for vaccination of the old in 2006. This vaccine induces antibody and T cell responses [81], and its clinical efficacy has been demonstrated in a large double-blind placebo-controlled study [82] in persons over 60 years of age. Compared to placebo the incidence of herpes zoster was reduced by 51.3 % (95 % CI 44.2–57.6) and the incidence of PHN by 66.5 % (95 % CI 44.5–79.2) in the vaccinated population. Whereas the protective effect against PHN was independent of age, the efficacy against herpes zoster was age-dependent and decreased to only 37.6 % in persons older than 69 years. The duration of the protection has been studied in follow-up studies demonstrating a decrease in vaccine efficacy with each year after vaccination, and reporting an estimated vaccine efficacy of 21.1 % (95 % CI 20.9–30.4) for the prevention of herpes zoster and 35.4 % (95 % CI 8.8–55.8 %) for PHN in years 7–10 [83]. These results raise the question of optimal age for vaccination. The vaccine is licensed for adults over the age of 50 years, but most countries recommend vaccination at age 60. A clinical trial evaluating the safety and immunogenicity of a second dose of vaccine 10 years later is currently being performed (ClinicalTrials.gov,

NCT01245751) and will provide information about whether a booster dose of herpes zoster vaccine is beneficial. Despite ACIP (Advisory Committee on Immunization Practices) recommendations having been in place for several years, the uptake of the vaccine remains low in the US, reaching approximately 20 % in managed care populations, and even less in the general population [84].

The incidence of herpes zoster is also high in immunocompromised patients, e.g. after transplantation, in cancer patients, and in HIV-positive individuals, as cell-mediated immunity to VZV is impaired in these patients. As the current herpes zoster vaccine contains live-attenuated virus, it cannot be used in immunocompromised patients due to safety issues [85]. A novel, inactivated vaccine against herpes zoster containing the viral glycoprotein E in combination with the liposome-based $AS01_B$ adjuvant system (MPL and QS21) has been developed, and phase I/II studies have demonstrated safety and immunogenicity in hematopoietic cell transplant recipients [86], in HIV-patients [87] and in older adults [88, 89]. Overall clinical efficacy against herpes zoster was 97.2 % (95 % CI 93.7–99.0, $p < 0.001$) in a phase III randomized placebo-controlled trial including more than 15,000 adults over the age of 50. Remarkably, vaccine efficacy in adults who were 70 years of age or older, was similar to the younger study cohorts [90]. This vaccine might replace the live-attenuated herpes zoster vaccine in the future depending on further data provided by ongoing clinical trials including information on long-term protection and the potential need for booster doses.

5.3   Vaccines Recommended for all Adults

5.3.1   Tetanus and Diphtheria Vaccine

Regular booster immunizations against tetanus and diphtheria throughout life using a combined vaccine are recommended in many countries [91]. *Clostridium tetani* is found ubiquitously in soil and infection occurs mainly through contaminated wounds, whereas *Corynebacterium diphtheria* is transmitted via droplets. Despite a low incidence rate in Europe with 161 cases of tetanus and 36 cases of diphtheria per year (mean from 2009 to 2014) [92], vaccination against these diseases is still of importance. Due to the mode of transmission, vaccination does not decrease the prevalence of the pathogen, and there is no herd immunity effect for tetanus. Every individual needs to be vaccinated in order to be protected. A large outbreak of diphtheria in the former Soviet Union in the early 1990s with more than 140,000 cases clearly demonstrated that the pathogen is still present and can spread rapidly in a partially unprotected population [93]. For one of our studies a cohort of persons above the age of 60 was recruited in Austria in order to receive a booster shot against tetanus and diphtheria. Efforts to retrieve information about vaccination history were only partially successful, as it was not possible to determine the exact time point of the last vaccination in 10 % and 53 % of the cases for tetanus or diphtheria, respectively [94]. Similar data are available from France and Belgium [95, 96]. Appropriate vaccination documentation is crucial to deliver booster vaccinations at the right time points. Vaccination against tetanus is recommended when wounds which could potentially be contaminated with soil (e.g. after accidents) are treated and no recent vaccination is documented. Official recommendations specify the use of a combined tetanus/diphtheria vaccine in such cases in order to avoid multiple tetanus shots and a lack of diphtheria vaccination, but in the past single tetanus vaccinations were common. In the above-mentioned study only 31 % of the participants had received their last tetanus/diphtheria vaccination as a combined vaccine [94].

Several studies have demonstrated that tetanus- and diphtheria-specific antibody concentrations are frequently below levels considered to be protective for adults, and particularly for the old [95–98]. In an Austrian cohort 12 % and 65 % of persons older than 60 years were not protected against tetanus or diphtheria, respectively [94]. The elapsed time since the last vaccination, as well as age, has an impact on antibody titers against tetanus. At all time points antibody concentrations are lower in the old compared to young adults and are negatively correlated with the elapsed time since the last vaccination [99]. Upon booster vaccination antibody concentrations increase in most older persons, but approximately 10 % do not develop sufficient antibody levels against diphtheria after a single booster shot [94]. This cohort was followed for 5 years after the booster shot and we showed that at this time point again, 10 % and 45 % of the participants were not protected against tetanus or diphtheria, respectively. A second booster shot was administered, as Austrian vaccine recommendations include shortened tetanus and diphtheria booster intervals of 5 years for persons over the age of 60. Similar to the first booster all participants developed protective antibody concentrations against tetanus, but 6 % again did not respond to the diphtheria vaccine [94]. In conclusion, single-booster shots later in life do not elicit sufficient and long-lasting antibody responses in a substantial portion of the old.

Due to the poor vaccination documentation for many older adults, it is difficult to reliably assess correct primary vaccination in childhood and the number of booster shots administered throughout life. A study in France showed that the number of vaccine doses received in life decreases with age. Young adults (<30 years) received on average 7.1 doses (95 % CI 6.9–7.2) doses of tetanus vaccine, which corresponds well with recommendations of five doses during childhood/adolescence and 10 year-booster intervals afterwards. However, persons aged 50–60 years received only 5.7 (95 % CI 4.6–6.8) doses over their life-time, indicating a lack of regular booster vaccinations [95]. Vaccination strategies for the future should include regular and well-documented booster shots throughout life, as post-booster antibody concentrations correlate with pre-booster antibody concentrations [98]. The success of primary vaccination late in life for persons without adequate priming remains to be elucidated, as the problem of memory generation late in life is well documented in animal models [100]. Considerations about improved vaccines, particularly for diphtheria, should also be taken into account.

5.3.2   Pertussis Vaccine

Paediatric vaccination against pertussis is widely used and accepted. Epidemiological data show that pertussis, which was considered to affect mainly very young children, is also relevant for older age groups. Surveillance for pertussis is difficult and therefore the real burden of disease is probably underestimated, but several studies document an increased incidence of pertussis in adults and particularly in the old [101], who experience higher morbidity and mortality [102]. These observations might partially be attributed to improved surveillance and awareness among physicians, but it has also been demonstrated that immunity against pertussis wanes both after natural infection and even faster after vaccination [103, 104]. Booster doses of combined tetanus/diphtheria/pertussis vaccine are well tolerated and immunogenic when given periodically to young or older adults [105], but it has to be considered that antibody concentrations 4 weeks after the vaccination are lower in the elderly compared with young adults [98]. Some countries recommend one dose of pertussis-containing vaccine during adulthood, but others implemented regular booster vaccination with a combined vaccine including pertussis antigens throughout life in their vaccination schedules (Table 5.1).

5.4   Novel Vaccines for the Older Adult

Currently, we still lack vaccines against many pathogens that are of particular relevance for the old, despite substantial effort put into their development. Respiratory syncytial virus (RSV) is a major cause of severe respiratory infection in infants. For adults, RSV infection is usually associated with only mild or moderate symptoms. However, persons with underlying chronic diseases and frail older adults are susceptible to severe disease. It has been estimated that almost 18,000 hospitalizations and 8400 deaths caused by RSV occur in the United Kingdom per season and that 79 % of the hospitalizations and 93 % of deaths were in persons older than 65 years. High-risk elderly (with chronic conditions such as COPD, cardiovascular disease, renal and liver disorders, immunosuppressive therapy etc.) have an increased rate of hospitalization (fourfold) and death (twofold) compared to low-risk older patients [106]. A first paediatric vaccine candidate against RSV has been developed in the 1960s, but was associated with a risk of enhanced disease in vaccinated children [107]. This failure hampered vaccine development for a long time, but nowadays several novel vaccine candidates against RSV are in clinical development, including vaccines based on recombinant proteins, virus-like particles and live-attenuated vaccines [108]. It will be of utmost importance that these vaccine candidates will be further developed not only for the paediatric market, but also tested in adults and the old.

Vaccines against nosocomial infections are highly desirable, particularly as antibiotic resistance is increasing for many bacterial species. Nevertheless, no such vaccine is currently licensed. The risk of nosocomial infections is particularly high for the old, as they are more often hospitalized, are at higher risk of invasive procedures (surgery, prostheses, catheters etc.) and are more susceptible to severe consequences and death in case of infection. *Staphylococcus aureus* infections range from mild skin infections to potentially fatal pneumonia and invasive disease [109]. Two vaccine candidates targeting single antigens have been clinically tested [110, 111], but had only limited or no efficacy against *S. aureus* infection. Three next-generation vaccines comprising several antigens are currently in early clinical development [112]. Most cases of nosocomial infectious diarrhea are caused by *Clostridium difficile* [113] and 81 % of cases in England have been reported to occur in older persons, in whom mortality is also increased [114]. Highly virulent strains have emerged and become endemic in hospitals, and antibiotic resistance is a growing problem. Clinical symptoms are caused by bacterial toxins and therefore vaccination strategies aim to elicit toxin-neutralizing antibodies. Clinical development is ongoing for three toxin-based vaccines and first results indicate the induction of neutralizing antibodies in healthy adults including older individuals [112]. Vaccines against these two and other nosocomial pathogens, such as *Escherichia coli*, *Klebsiella pneumoniae* and *Candida ssp.* [115] have the potential to save many lives and to substantially reduce healthcare costs.

5.5 Conclusions

Immunosenescence is responsible for impaired immune responses after vaccination in older individuals leading to decreased clinical efficacy of many vaccines in this age group. Nevertheless, vaccination is still the most effective measure to prevent infectious diseases. Childhood vaccination programs are well-established and accepted world-wide, but public awareness for adult vaccination is often limited. Most countries implement an adult vaccination schedule including specific recommendations for the old, but coverage is frequently poor [10, 116]. Recommendations, particularly for new vaccines, differ greatly between European countries [117] and a unified European recommendation including all age groups would be desirable. Basic research is essential to understand mechanisms underlying immunosenescence. Cell-intrinsic defects of individual cell types but also the complex interplay of cells and soluble factors need to be studied in order to fully understand age-related changes of the immune system. This knowledge will enable us to rationally design new vaccines that specifically target the aged immune system and might be able to overcome its limitations. It is of crucial importance that older individuals are included in clinical studies testing vaccines intended to be used in this age group. Most clinical trials, particularly in early stages, are currently only carried out in healthy young adults.

References

1. Thompson WW, Shay DK, Weintraub E, Brammer L, Cox N, Anderson LJ, et al. Mortality associated with influenza and respiratory syncytial virus in the United States. JAMA. 2003;289:179–86.
2. Thompson WW, Shay DK, Weintraub E, Brammer L, Bridges CB, Cox NJ, et al. Influenza-associated hospitalizations in the United States. JAMA. 2004;292:1333–40.
3. Pinchinat S, Cebrian-Cuenca AM, Bricout H, Johnson RW. Similar herpes zoster incidence across Europe: results from a systematic literature review. BMC Infect Dis. 2013;13:170.
4. Thigpen MC, Whitney CG, Messonnier NE, Zell ER, Lynfield R, Hadler JL, et al. Bacterial meningitis in the United States, 1998-2007. N Engl J Med. 2011;364:2016–25.
5. Janssens JP, Krause KH. Pneumonia in the very old. Lancet Infect Dis. 2004;4:112–24.

6. Edwards MS, Baker CJ. Group B streptococcal infections in elderly adults. Clin Infect Dis. 2005;41:839–47.
7. Kish TD, Chang MH, Fung HB. Treatment of skin and soft tissue infections in the elderly: a review. Am J Geriatr Pharmacother. 2010;8:485–513.
8. Matthews SJ, Lancaster JW. Urinary tract infections in the elderly population. Am J Geriatr Pharmacother. 2011;9:286–309.
9. Weinberger B, Herndler-Brandstetter D, Schwanninger A, Weiskopf D, Grubeck-Loebenstein B. Biology of immune responses to vaccines in elderly persons. Clin Infect Dis. 2008;46:1078–84.
10. Mereckiene, J. Seasonal influenza vaccination in Europe. 2015 http://ecdc.europa.eu/en/publications/Publications/Seasonal-influenza-vaccination-Europe-2012-13.pdf. Accessed 21 Dec 2015.
11. Belshe RB, Coelingh K, Ambrose CS, Woo JC, Wu X. Efficacy of live attenuated influenza vaccine in children against influenza B viruses by lineage and antigenic similarity. Vaccine. 2010;28:2149–56.
12. Kieninger D, Sheldon E, Lin WY, Yu CJ, Bayas JM, Gabor JJ, et al. Immunogenicity, reactogenicity and safety of an inactivated quadrivalent influenza vaccine candidate versus inactivated trivalent influenza vaccine: a phase III, randomized trial in adults aged >/=18 years. BMC Infect Dis. 2013;13:343.
13. McKeage K. Inactivated quadrivalent split-virus seasonal influenza vaccine (Fluarix(R) quadrivalent): a review of its use in the prevention of disease caused by influenza A and B. Drugs. 2013;73:1587–94.
14. Brydak LB, Machala M, Mysliwska J, Mysliwski A, Trzonkowski P. Immune response to influenza vaccination in an elderly population. J Clin Immunol. 2003;23:214–22.
15. Stepanova L, Naykhin A, Kolmskog C, Jonson G, Barantceva I, Bichurina M, et al. The humoral response to live and inactivated influenza vaccines administered alone and in combination to young adults and elderly. J Clin Virol. 2002;24:193–201.
16. Mysliwska J, Trzonkowski P, Szmit E, Brydak LB, Machala M, Mysliwski A. Immunomodulating effect of influenza vaccination in the elderly differing in health status. Exp Gerontol. 2004;39:1447–58.
17. Goodwin K, Viboud C, Simonsen L. Antibody response to influenza vaccination in the elderly: a quantitative review. Vaccine. 2006;24:1159–69.
18. Gravenstein S, Drinka P, Duthie EH, Miller BA, Brown CS, Hensley M, et al. Efficacy of an influenza hemagglutinin-diphtheria toxoid conjugate vaccine in elderly nursing home subjects during an influenza outbreak. J Am Geriatr Soc. 1994;42:245–51.
19. Wu YC, Kipling D, Dunn-Walters DK. Age-related changes in human peripheral blood IGH repertoire following vaccination. Front Immunol. 2012;3:193.
20. Jiang N, He J, Weinstein JA, Penland L, Sasaki S, He XS, et al. Lineage structure of the human antibody repertoire in response to influenza vaccination. Sci Transl Med. 2013;5:171ra19.
21. Sasaki S, Sullivan M, Narvaez CF, Holmes TH, Furman D, Zheng NY, et al. Limited efficacy of inactivated influenza vaccine in elderly individuals is associated with decreased production of vaccine-specific antibodies. J Clin Invest. 2011;121:3109–19.
22. Murasko DM, Bernstein ED, Gardner EM, Gross P, Munk G, Dran S, et al. Role of humoral and cell-mediated immunity in protection from influenza disease after immunization of healthy elderly. Exp Gerontol. 2002;37:427–39.
23. Zhou X, McElhaney JE. Age-related changes in memory and effector T cells responding to influenza A/H3N2 and pandemic A/H1N1 strains in humans. Vaccine. 2011;29:2169–77.
24. Rockwood K, Song X, MacKnight C, Bergman H, Hogan DB, McDowell I, et al. A global clinical measure of fitness and frailty in elderly people. CMAJ. 2005;173:489–95.
25. Yao X, Hamilton RG, Weng NP, Xue QL, Bream JH, Li H, et al. Frailty is associated with impairment of vaccine-induced antibody response and increase in post-vaccination influenza infection in community-dwelling older adults. Vaccine. 2011;29:5015–21.

26. Carrat F, Flahault A. Influenza vaccine: the challenge of antigenic drift. Vaccine. 2007;25:6852–62.
27. Call SA, Vollenweider MA, Hornung CA, Simel DL, McKinney WP. Does this patient have influenza? JAMA. 2005;293:987–97.
28. Boivin G, Hardy I, Tellier G, Maziade J. Predicting influenza infections during epidemics with use of a clinical case definition. Clin Infect Dis. 2000;31:1166–9.
29. Babcock HM, Merz LR, Fraser VJ. Is influenza an influenza-like illness? Clinical presentation of influenza in hospitalized patients. Infect Control Hosp Epidemiol. 2006;27:266–70.
30. Flamaing J, Engelmann I, Joosten E, van Ranst M, Verhaegen J, Peetermans WE. Viral lower respiratory tract infection in the elderly: a prospective in-hospital study. Eur J Clin Microbiol Infect Dis. 2003;22:720–5.
31. Talbot HK, Keitel W, Cate TR, Treanor J, Campbell J, Brady RC, et al. Immunogenicity, safety and consistency of new trivalent inactivated influenza vaccine. Vaccine. 2008;26:4057–61.
32. Govaert TM, Thijs CT, Masurel N, Sprenger MJ, Dinant GJ, Knottnerus JA. The efficacy of influenza vaccination in elderly individuals. A randomized double-blind placebo-controlled trial. JAMA. 1994;272:1661–5.
33. Praditsuwan R, Assantachai P, Wasi C, Puthavatana P, Kositanont U. The efficacy and effectiveness of influenza vaccination among Thai elderly persons living in the community. J Med Assoc Thai. 2005;88:256–64.
34. Allsup S, Haycox A, Regan M, Gosney M. Is influenza vaccination cost effective for healthy people between ages 65 and 74 years? A randomised controlled trial. Vaccine. 2004;23:639–45.
35. Beyer WE, McElhaney J, Smith DJ, Monto AS, Nguyen-Van-Tam JS, Osterhaus AD. Cochrane re-arranged: support for policies to vaccinate elderly people against influenza. Vaccine. 2013;31:6030–3.
36. Osterholm MT, Kelley NS, Sommer A, Belongia EA. Efficacy and effectiveness of influenza vaccines: a systematic review and meta-analysis. Lancet Infect Dis. 2012;12:36–44.
37. Rivetti D, Jefferson T, Thomas R, Rudin M, Rivetti A, Di PC, et al. Vaccines for preventing influenza in the elderly. Cochrane Database Syst Rev. 2006;3, CD004876.
38. Cools HJ, Gussekloo J, Remmerswaal JE, Remarque EJ, Kroes AC. Benefits of increasing the dose of influenza vaccine in residents of long-term care facilities: a randomized placebo-controlled trial. J Med Virol. 2009;81:908–14.
39. Chen WH, Cross AS, Edelman R, Sztein MB, Blackwelder WC, Pasetti MF. Antibody and Th1-type cell-mediated immune responses in elderly and young adults immunized with the standard or a high dose influenza vaccine. Vaccine. 2011;29:2865–73.
40. Nace DA, Lin CJ, Ross TM, Saracco S, Churilla RM, Zimmerman RK. Randomized, controlled trial of high-dose influenza vaccine among frail residents of long-term care facilities. J Infect Dis. 2015;211:1915–24.
41. Richardson DM, Medvedeva EL, Roberts CB, Linkin DR. Comparative effectiveness of high-dose versus standard-dose influenza vaccination in community-dwelling veterans. Clin Infect Dis. 2015;61:171–6.
42. Izurieta HS, Thadani N, Shay DK, Lu Y, Maurer A, Foppa IM, et al. Comparative effectiveness of high-dose versus standard-dose influenza vaccines in US residents aged 65 years and older from 2012 to 2013 using Medicare data: a retrospective cohort analysis. Lancet Infect Dis. 2015;15:293–300.
43. DiazGranados CA, Dunning AJ, Kimmel M, Kirby D, Treanor J, Collins A, et al. Efficacy of high-dose versus standard-dose influenza vaccine in older adults. N Engl J Med. 2014;371:635–45.
44. Nicolas JF, Guy B. Intradermal, epidermal and transcutaneous vaccination: from immunology to clinical practice. Expert Rev Vaccines. 2008;7:1201–14.
45. Durando P, Iudici R, Alicino C, Alberti M, de Florentis D, Ansaldi, F et al. Adjuvants and alternative routes of administration towards the development of the ideal influenza vaccine. Hum Vaccin. 2011;7(Suppl):29–40.

46. Holland D, Booy R, de Looze F, Eizenberg P, McDonald J, Karrasch J, et al. Intradermal influenza vaccine administered using a new microinjection system produces superior immunogenicity in elderly adults: a randomized controlled trial. J Infect Dis. 2008;198:650–8.

47. Ott G, Barchfeld GL, Van Nest G. Enhancement of humoral response against human influenza vaccine with the simple submicron oil/water emulsion adjuvant MF59. Vaccine. 1995;13:1557–62.

48. O'Hagan DT, Ott GS, DeGregorio E, Seubert A. The mechanism of action of MF59—an innately attractive adjuvant formulation. Vaccine. 2012;30:4341–8.

49. Beyer WE, Nauta JJ, Palache AM, Giezeman KM, Osterhaus AD. Immunogenicity and safety of inactivated influenza vaccines in primed populations: a systematic literature review and meta-analysis. Vaccine. 2011;29:5785–92.

50. Banzhoff A, Nacci P, Podda A. A new MF59-adjuvanted influenza vaccine enhances the immune response in the elderly with chronic diseases: results from an immunogenicity meta-analysis. Gerontology. 2003;49:177–84.

51. Mannino S, Villa M, Apolone G, Weiss NS, Groth N, Aquino I, et al. Effectiveness of adjuvanted influenza vaccination in elderly subjects in northern Italy. Am J Epidemiol. 2012;176:527–33.

52. Iob A, Brianti G, Zamparo E, Gallo T. Evidence of increased clinical protection of an MF59-adjuvant influenza vaccine compared to a non-adjuvant vaccine among elderly residents of long-term care facilities in Italy. Epidemiol Infect. 2005;133:687–93.

53. Legrand J, Vergu E, Flahault A. Real-time monitoring of the influenza vaccine field effectiveness. Vaccine. 2006;24:6605–11.

54. Ansaldi F, Bacilieri S, Durando P, Sticchi L, Valle L, Montomoli E, et al. Cross-protection by MF59-adjuvanted influenza vaccine: neutralizing and haemagglutination-inhibiting antibody activity against A(H3N2) drifted influenza viruses. Vaccine. 2008;26:1525–9.

55. DelGiudice G, Hilbert AK, Bugarini R, Minutello A, Popova O, Toneatto D, et al. An MF59-adjuvanted inactivated influenza vaccine containing A/Panama/1999 (H3N2) induced broader serological protection against heterovariant influenza virus strain A/Fujian/2002 than a subunit and a split influenza vaccine. Vaccine. 2006;24:3063–5.

56. Baldo V, Baldovin T, Pellegrini M, Angiolelli G, Majori S, Floreani A, et al. Immunogenicity of three different influenza vaccines against homologous and heterologous strains in nursing home elderly residents. Clin Dev Immunol. 2010;2010:517198.

57. O'Hagan DT, Rappuoli R, DeGregorio E, Tsai T, Del GG. MF59 adjuvant: the best insurance against influenza strain diversity. Expert Rev Vaccines. 2011;10:447–62.

58. Del Giudice G, Rappuoli R. Inactivated and adjuvanted influenza vaccines. Curr Top Microbiol Immunol. 2015;386:151–80.

59. Klucker MF, Dalencon F, Probeck P, Haensler J. AF03, an alternative squalene emulsion-based vaccine adjuvant prepared by a phase inversion temperature method. J Pharm Sci. 2012;101:4490–500.

60. Garcon N, Vaughn DW, Didierlaurent AM. Development and evaluation of AS03, an Adjuvant System containing alpha-tocopherol and squalene in an oil-in-water emulsion. Expert Rev Vaccines. 2012;11:349–66.

61. Moser C, Amacker M, Zurbriggen R. Influenza virosomes as a vaccine adjuvant and carrier system. Expert Rev Vaccines. 2011;10:437–46.

62. Wiersma LC, Rimmelzwaan GF, de Vries RD. Developing universal influenza vaccines: hitting the nail, not just on the head. Vaccines (Basel). 2015;3:239–62.

63. Prevention of pneumococcal disease: recommendations of the Advisory Committee on Immunization Practices (ACIP). MMWR Recomm Rep. 1997;46: 1–24.

64. Moberley SA, Holden J, Tatham DP, Andrews RM. Vaccines for preventing pneumococcal infection in adults. Cochrane Database Syst Rev. 2008;1, CD000422.

65. Vila-Corcoles A, Salsench E, Rodriguez-Blanco T, Ochoa-Gondar O, de Diego C, Valdivieso A, et al. Clinical effectiveness of 23-valent pneumococcal polysaccharide vaccine against pneumonia in middle-aged and older adults: a matched case-control study. Vaccine. 2009;27:1504–10.

66. Maruyama T, Taguchi O, Niederman MS, Morser J, Kobayashi H, Kobayashi T, et al. Efficacy of 23-valent pneumococcal vaccine in preventing pneumonia and improving survival in nursing home residents: double blind, randomised and placebo controlled trial. BMJ. 2010;340:c1004.

67. Melegaro A, Edmunds WJ. The 23-valent pneumococcal polysaccharide vaccine. Part I. Efficacy of PPV in the elderly: a comparison of meta-analyses. Eur J Epidemiol. 2004;19:353–63.

68. Feldman C, Anderson R. Review: current and new generation pneumococcal vaccines. J Infect. 2014;69:309–25.

69. Kumar R, Burns EA. Age-related decline in immunity: implications for vaccine responsiveness. Expert Rev Vaccines. 2008;7:467–79.

70. De Donato S, Granoff D, Minutello M, Lecchi G, Faccini M, Agnello M, et al. Safety and immunogenicity of MF59-adjuvanted influenza vaccine in the elderly. Vaccine. 1999;17:3094–101.

71. Lexau CA, Lynfield R, Danila R, Pilishvili T, Facklam R, Farley MM, et al. Changing epidemiology of invasive pneumococcal disease among older adults in the era of pediatric pneumococcal conjugate vaccine. JAMA. 2005;294:2043–51.

72. Jackson LA, Gurtman A, Vancleeff M, Jansen KU, Jayawardene D, Devlin C, et al. Immunogenicity and safety of a 13-valent pneumococcal conjugate vaccine compared to a 23-valent pneumococcal polysaccharide vaccine in pneumococcal vaccine-naive adults. Vaccine. 2013;31:3577–84.

73. Bonten MJ, Huijts SM, Bolkenbaas M, Webber C, Patterson S, Gault S, et al. Polysaccharide conjugate vaccine against pneumococcal pneumonia in adults. N Engl J Med. 2015;372:1114–25.

74. van Werkhoven CH, Huijts SM, Bolkenbaas M, Grobbee DE, Bonten MJ. The impact of Age on the efficacy of 13-valent pneumococcal conjugate vaccine in elderly. Clin Infect Dis. 2015;61:1835–8.

75. McFetridge R, Meulen AS, Folkerth SD, Hoekstra JA, Dallas M, Hoover PA, et al. Safety, tolerability, and immunogenicity of 15-valent pneumococcal conjugate vaccine in healthy adults. Vaccine. 2015;33:2793–9.

76. Dagan R. Serotype replacement in perspective. Vaccine. 2009;27 Suppl 3:C22–4.

77. Esposito S, Principi N. Direct and indirect effects of the 13-valent pneumococcal conjugate vaccine administered to infants and young children. Future Microbiol. 2015;10:1599–607.

78. Burke BL, Steele RW, Beard OW, Wood JS, Cain TD, Marmer DJ. Immune responses to varicella-zoster in the aged. Arch Intern Med. 1982;142:291–3.

79. Katz J, Cooper EM, Walther RR, Sweeney EW, Dworkin RH. Acute pain in herpes zoster and its impact on health-related quality of life. Clin Infect Dis. 2004;39:342–8.

80. Scott FT, Johnson RW, Leedham-Green M, Davies E, Edmunds WJ, Breuer J. The burden of Herpes Zoster: a prospective population based study. Vaccine. 2006;24:1308–14.

81. Levin MJ, Oxman MN, Zhang JH, Johnson GR, Stanley H, Hayward AR, et al. Varicella-zoster virus-specific immune responses in elderly recipients of a herpes zoster vaccine. J Infect Dis. 2008;197:825–35.

82. Oxman MN, Levin MJ, Johnson GR, Schmader KE, Straus SE, Gelb LD, et al. A vaccine to prevent herpes zoster and postherpetic neuralgia in older adults. N Engl J Med. 2005;352:2271–84.

83. Morrison VA, Johnson GR, Schmader KE, Levin MJ, Zhang JH, Looney DJ, et al. Long-term persistence of zoster vaccine efficacy. Clin Infect Dis. 2015;60:900–9.

84. Hechter RC, Tartof SY, Jacobsen SJ, Smith N, Tseng HF. Trends and disparity in zoster vaccine uptake in a managed care population. Vaccine. 2013;31:4564–8.

85. Hales CM, Harpaz R, Ortega-Sanchez I, Bialek SR. Update on recommendations for use of herpes zoster vaccine. MMWR Morb Mortal Wkly Rep. 2014;63:729–31.

86. Stadtmauer EA, Sullivan KM, Marty FM, Dadwal SS, Papanicolaou GA, Shea TC, et al. A phase 1/2 study of an adjuvanted varicella-zoster virus subunit vaccine in autologous hematopoietic cell transplant recipients. Blood. 2014;124:2921–9.

87. Berkowitz EM, Moyle G, Stellbrink HJ, Schurmann D, Kegg S, Stoll M, et al. Safety and immunogenicity of an adjuvanted herpes zoster subunit candidate vaccine in HIV-infected adults: a phase 1/2a randomized, placebo-controlled study. J Infect Dis. 2015;211:1279–87.
88. Leroux-Roels I, Leroux-Roels G, Clement F, Vandepapeliere P, Vassilev V, Ledent E, et al. A phase 1/2 clinical trial evaluating safety and immunogenicity of a varicella zoster glycoprotein e subunit vaccine candidate in young and older adults. J Infect Dis. 2012;206:1280–90.
89. Chlibek R, Bayas JM, Collins H, de la Pinta ML, Ledent E, Mols JF, et al. Safety and immunogenicity of an AS01-adjuvanted varicella-zoster virus subunit candidate vaccine against herpes zoster in adults >=50 years of age. J Infect Dis. 2013;208:1953–61.
90. Lal H, Cunningham AL, Godeaux O, Chlibek R, Diez-Domingo J, Hwang SJ, et al. Efficacy of an adjuvanted herpes zoster subunit vaccine in older adults. N Engl J Med. 2015;372:2087–96.
91. Weinberger B. Adult vaccination against tetanus and diphtheria: the European perspective. Clin Exp Immunol. 2016. doi:10.1111/cei.12822.
92. Vaccine-preventable diseases: monitoring system. 2016 http://apps.who.int/immunization_monitoring/globalsummary. Accessed 14 Jan 2016.
93. Vitek CR, Wharton M. Diphtheria in the former Soviet Union: reemergence of a pandemic disease. Emerg Infect Dis. 1998;4:539–50.
94. Weinberger B, Schirmer M, Matteucci GR, Siebert U, Fuchs D, Grubeck-Loebenstein B. Recall responses to tetanus and diphtheria vaccination are frequently insufficient in elderly persons. PLoS One. 2013;8, e82967.
95. Launay O, Toneatti C, Bernede C, Njamkepo E, Petitprez K, Leblond A, et al. Antibodies to tetanus, diphtheria and pertussis among healthy adults vaccinated according to the French vaccination recommendations. Hum Vaccin. 2009;5:341–6.
96. Van Damme P, Burgess M. Immunogenicity of a combined diphtheria-tetanus-acellular pertussis vaccine in adults. Vaccine. 2004;22:305–8.
97. Steger MM, Maczek C, Berger P, Grubeck-Loebenstein B. Vaccination against tetanus in the elderly: do recommended vaccination strategies give sufficient protection. Lancet. 1996;348:762.
98. Kaml M, Weiskirchner I, Keller M, Luft T, Hoster E, Hasford J, et al. Booster vaccination in the elderly: Their success depends on the vaccine type applied earlier in life as well as on pre-vaccination antibody titers. Vaccine. 2006;24:6808–11.
99. Hainz U, Jenewein B, Asch E, Pfeiffer KP, Berger P, Grubeck-Loebenstein B. Insufficient protection for healthy elderly adults by tetanus and TBE vaccines. Vaccine. 2005;23:3232–5.
100. Haynes L, Eaton SM, Burns EM, Randall TD, Swain SL. CD4 T cell memory derived from young naive cells functions well into old age, but memory generated from aged naive cells functions poorly. Proc Natl Acad Sci U S A. 2003;100:15053–8.
101. Ridda I, Yin JK, King C, Raina MC, McIntyre P. The importance of pertussis in older adults: a growing case for reviewing vaccination strategy in the elderly. Vaccine. 2012;30:6745–52.
102. Gil A, Oyaguez I, Carrasco P, Gonzalez A. Hospital admissions for pertussis in Spain, 1995-1998. Vaccine. 2001;19:4791–4.
103. Barreto L, Guasparini R, Meekison W, Noya F, Young L, Mills E. Humoral immunity 5 years after booster immunization with an adolescent and adult formulation combined tetanus, diphtheria, and 5-component acellular pertussis vaccine. Vaccine. 2007;25:8172–9.
104. Wendelboe AM, van Rie A, Salmaso S, Englund JA. Duration of immunity against pertussis after natural infection or vaccination. Pediatr Infect Dis J. 2005;24:S58–61.
105. Halperin SA, Scheifele D, de Serres G, Noya F, Meekison W, Zickler P, et al. Immune responses in adults to revaccination with a tetanus toxoid, reduced diphtheria toxoid, and acellular pertussis vaccine 10 years after a previous dose. Vaccine. 2012;30:974–82.
106. Fleming DM, Taylor RJ, Lustig RL, Schuck-Paim C, Haguinet F, Webb DJ, et al. Modelling estimates of the burden of Respiratory Syncytial virus infection in adults and the elderly in the United Kingdom. BMC Infect Dis. 2015;15:443.

107. Kim HW, Canchola JG, Brandt CD, Pyles G, Chanock RM, Jensen K, et al. Respiratory syncytial virus disease in infants despite prior administration of antigenic inactivated vaccine. Am J Epidemiol. 1969;89:422–34.
108. Shaw CA, Ciarlet M, Cooper BW, Dionigi L, Keith P, O'Brien KB, et al. The path to an RSV vaccine. Curr Opin Virol. 2013;3:332–42.
109. Lowy FD. Staphylococcus aureus infections. N Engl J Med. 1998;339:520–32.
110. Shinefield H, Black S, Fattom A, Horwith G, Rasgon S, Ordonez J, et al. Use of a Staphylococcus aureus conjugate vaccine in patients receiving hemodialysis. N Engl J Med. 2002;346:491–6.
111. Fowler VG, Allen KB, Moreira ED, Moustafa M, Isgro F, Boucher HW, et al. Effect of an investigational vaccine for preventing Staphylococcus aureus infections after cardiothoracic surgery: a randomized trial. JAMA. 2013;309:1368–78.
112. Swanson KA, Schmitt HJ, Jansen KU, Anderson AS. Adult vaccination. Hum Vaccin Immunother. 2015;11:150–5.
113. Simor AE, Bradley SF, Strausbaugh LJ, Crossley K, Nicolle LE. Clostridium difficile in long-term-care facilities for the elderly. Infect Control Hosp Epidemiol. 2002;23:696–703.
114. Karas JA, Enoch DA, Aliyu SH. A review of mortality due to Clostridium difficile infection. J Infect. 2010;61:1–8.
115. Cross AS, Chen WH, Levine MM. A case for immunization against nosocomial infections. J Leukoc Biol. 2008;83:483–8.
116. Williams WW, Lu PJ, O'Halloran A, Bridges CB, Pilishvili T, Hales CM, et al. Noninfluenza vaccination coverage among adults—United States, 2012. MMWR Morb Mortal Wkly Rep. 2014;63:95–102.
117. Kanitz EE, Wu LA, Giambi C, Strikas RA, Levy-Bruhl D, Stefanoff P, et al. Variation in adult vaccination policies across Europe: an overview from VENICE network on vaccine recommendations, funding and coverage. Vaccine. 2012;30:5222–8.

Immunosenescence and the Ageing Lung

6

Krisztian Kvell and Judit E. Pongracz

Abstract

Ageing is generally defined as the progressive decline of homeostasis that occurs after the reproductive phase of life is complete and the *"soma becomes disposable"* and death is inevitable according to one theory of ageing. The complexity of the ageing process becomes strikingly evident in the lung where tissue maintenance and repair suffer from damage at the genetic level as well as tissue level. Moreover, lung function declines steadily in adulthood and if data for older adults are extrapolated, the outcome suggests an upper age limit beyond which life becomes impossible. In this chapter we cover the main changes to lung structure and function with age and the impact on respiratory health. We also describe the role that an aged immune system may play in the age-related decline in lung function and the major involvement of altered signalling through developmental pathways with special focus on PPARγ.

Keywords

Lung function • Inflammation • PPARγ • Wnt • Ageing

6.1 Introduction

In a recent article [1] the association of age with lung function decline was summarized and showed a linear decline from maturity. If trajectories depicted in the paper are extrapolated (Fig. 6.1), it becomes evident that the absolute extent of human life is limited, at least in part, by respiratory function to about 130 years but currently there are no confirmed cases available of people who lived to the absolute limit of pulmonary functional decline. A rare exception is Mrs Tuti Yusupova of Uzbekistan who died in 2014 apparently at the age of 134 [2]. Although caution is

K. Kvell • J.E. Pongracz (✉)

Department of Pharmaceutical Biotechnology, School of Pharmacy, and Szentagothai Research Center, University of Pecs, 2 Rokus Str, Pecs 7624, Hungary

e-mail: pongracz.e.judit@pte.hu

© Springer International Publishing Switzerland 2017

V. Bueno et al. (eds.), *The Ageing Immune System and Health*,

DOI 10.1007/978-3-319-43365-3_6

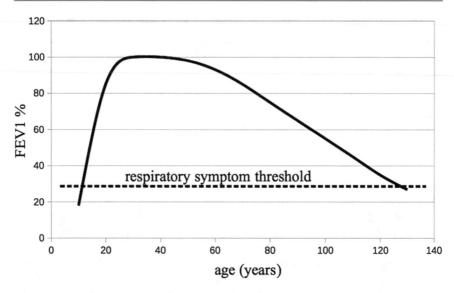

Fig. 6.1 Correlation of lung capacity with age (*dotted line* shows predicted respiratory symptom threshold), using data from [1]

required as her age has not been authenticated, she died as a very old person approaching the absolute limits of human life. The fully authenticated age to which any human has ever lived is 122 years and 164 days by Jeanne Louise Calment of France [3], who was born in 1875 and died in 1997. Ms Calment's and Ms Yusupova's very old age suggests that they suffered no detrimental co-morbidities and their genetic make-up regulating lung development, function and immune regulation was an enviably perfect combination. To understand more representative human ageing we have to use data from large population studies that have measured many aspects of human physiology, including lung function, over the lifespan.

6.2   Structural Changes of the Lungs During Pulmonary Senescence

The aged lung is characterised by airspace enlargement similar to, but not identical with acquired emphysema [4]. Such tissue damage is detected even in non-smokers above 50 years of age as the septa of the lung alveoli are destroyed and the enlarged alveolar structures result in a decreased surface for gas exchange [5] (Fig. 6.2). Tobacco smoking, pollution and hereditary factors are all involved in the regulation of emphysema making it difficult to separate the effect of accumulating environmental factors from the process of physiological ageing. Nevertheless, as total tissue mass and the number of capillaries decrease and formation of new alveoli becomes limited breathing difficulty is inevitable. Additional problems are that surfactant production decreases with age [6] increasing the effort needed to expand the lungs

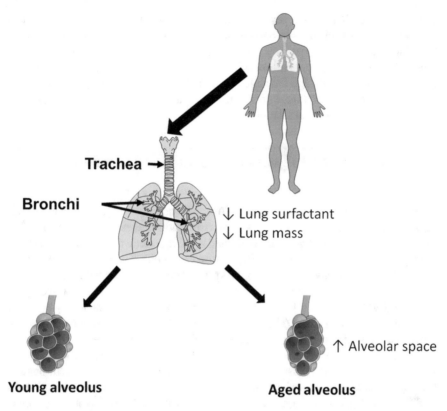

Fig. 6.2 Age-related changes of in the lung alveoli. The aged lung is characterised by airspace enlargement as the septa of the lung alveoli are destroyed and the enlarged alveolar structures result in a decreased surface for gas exchange. Total lung tissue mass and the number of capillaries decrease with age and formation of new alveoli becomes limited. Additionally surfactant production decreases with age increasing the effort needed to expand the lungs during inhalation

during inhalation in the already reduced thoracic cavity volume where the weakened muscles are unable to thoroughly ventilate. It is therefore no wonder that the decline of lung capacity is detected with age even if no specific disease is diagnosed [7, 8].

As ageing is associated with respiratory muscle strength reduction, coughing becomes difficult making it progressively challenging to eliminate inhaled particles, pollens, microbes, etc. Additionally, ciliary beat frequency (CBF) slows down with age impairing the lungs' first line of defence: mucociliary clearance [9] as the cilia can no longer repel invading microorganisms and particles. Consequently e.g. bacteria can more easily colonise the airways leading to infections that are frequent in the pulmonary tract of the older adult. In contrast to CBF, mucus production increases with age. Having mucus present in the airways intuitively appears to be beneficial as at baseline the mucus layer is needed to trap and eliminate inhaled particles and to prevent desiccation of airway surfaces. Mucus consists of an assortment of mucins and these high molecular-weight glycoproteins provide viscoelastic, gel-forming and some anti-microbial properties to mucus. Understandably, the precise mucin composition of the mucus is important and whether it changes with age would be of high interest. Currently, however, very little is known about this aspect of mucin secretion. Interestingly, some recent animal studies revealed that while the production of mucus increases with age the ability of bronchial and alveolar cells to effectively produce mucus upon stimuli declines in the older animal [10] making the lungs more vulnerable to environmental factors.

6.3   Genetic and Epigenetic Regulation of Pulmonary Function

To identify the genetic background to lung function decline with age, several thousands of people have participated in various gene association studies [11, 12] that aimed to find correlations amongst gene locations, and specific genetic predisposition to senile emphysema as well as a closely associated disease: COPD. The result of such studies finally identified a locus on chromosome 4q31 that was associated with the percent of expected FEV1/FVC (as forced expiratory volume in 1 s [FEV1] and forced vital capacity [FVC]) ratio [11]. The locus was located in an intergenic region upstream of hedgehog interacting protein (HHIP), a hedgehog pathway gene with a known role in development. Additionally, six new genetic loci were identified that appeared to be associated with pulmonary function. These were located on chromosomes 2, 4, 5, 6 and 15 near genes including Tensin1 (TSN1) (2q35), encoding an actin-filament binding protein, four additional genes including Glutathione S-Transferase, C-terminal Domain (GSTCD) (4q24), Human Serotonin Receptor (HTR4) (5q32–33) a serotonin pathway gene, immune function genes Advanced Glycosylation End Product-Specific Receptor (AGER) and Palmitoyl-Protein Thioesterase 2 (PPT2) (6p21), a G-protein coupled receptor GPR126 (6q24.1) as well as Thrombospondin type 1 domain containing 4 gene (THSD4) (15q33) in the thrombospondin gene family [11]. Currently, it is not entirely clear how the elderly will directly benefit from the results of genome wide association studies. Certainly, further work is needed to identify the precise relevance to ageing and potential targets that could prolong pulmonary function.

   In addition to specific genes, pulmonary senescence is also regulated by heritable modifications in gene expression that is not coded in the DNA sequence itself, but is rather governed by post-translational modifications in histone proteins and DNA. These modifications include chromatin remodeling (histone acetylation, methylation, ubiquitination, phosphorylation, and sumoylation) and DNA methylation. One of the most investigated epigenetic modulators of the ageing process are class I histone deacetylases [13]. Apart from various other cellular functions, for example, histone deacetylase 2 (HDAC2) regulates glucocorticoid function in inhibiting inflammatory responses and protects against DNA damage and cellular senescence as well as premature ageing in response to oxidative stress. Unfortunately, in ageing COPD patients HDAC2 is post-translationally modified by cigarette smoke leading to its reduction via an ubiquitination-proteasome dependent degradation process rendering glucocorticoid containing anti-inflammatory drugs ineffective during their treatment

[14]. Recently, NAD$^+$-dependent deacetylases known as sirtuins (SIRT1–SIRT7) have also been widely investigated for their role in the regulation of the ageing process [15]. The best characterized is SIRT1 and although its function in prolonging lifespan is currently under debate, it has been shown that through deacetylation of many transcriptional factors, SIRT1 modulates key events in ageing [16] including the oxidative stress response, endothelial dysfunction, and inflammation [17, 18]. Whether sirtuins modulate lung function is currently under investigation.

Age-associated alterations in gene expression are also intensively investigated at the level of small "non-coding" micro-RNAs (miRNAs) that post-transcriptionally regulate gene expression. Several miRNAs have already been reported to regulate the expression of SIRT1 including miR-217 [19] in endothelial cells, a downstream target of p53 microRNA, miR34a [20, 21], as well as miR-199a and miR-132 that mediate the regulation of chemokine production [22] or HIF-1α function [23]. Recently, some miRNAs including miR-1, miR-122 and miR-375, miR-21, miR-206, miR-30a that regulate conserved pathways of ageing, including insulin/IGF signalling (IIS), DAF-12 signalling and mTOR signalling, have been linked to human age-related disorders such as heart-, muscle- and neurodegenerative diseases [24]. Whether they are also involved in pulmonary senescence is currently unknown. The constantly present low level inflammation characteristic of ageing is also regulated by microRNAs e.g. miR-146a and thus could contribute to lung functional decline, this microRNA is also an important regulator of toll-like receptor dependent signalling pathways and therefore epithelial cell dependent immune responses [25].

6.4 Inflammageing and Immune Responses in the Lung

Structural changes of the ageing lung are regulated by genetically coded and acquired qualities that are tightly interconnected with systemic immune dysfunction and chronic, low level systemic inflammation, termed inflammageing. Inflammageing is the basal activation of the innate immune system in the absence of an immunologic threat [26] and is marked by elevated levels of tissue and circulating pro-inflammatory cytokines including interleukin (IL)-1β, IL-6, and tumour necrosis factor-α (TNF-α). Inflammageing combined with blunted innate and adaptive immune responses (see Chaps. 1 and 2) affects the lungs ability to fight infections and also leads to tissue remodelling.

In the adult healthy lung several immune cell types are resident at various anatomical sites including bronchial, interstitial and alveolar macrophages, dendritic cells, interstitial T- and B-lymphocytes [27]. Their presence is highly important as our lungs are in constant and direct contact with the environment via epithelial surfaces. During ventilation the lung surfaces are exposed to various microbes, particles and potentially damaging physical forces. Together with a variety of pulmonary epithelial cell types, the resident macrophages, T-cells and dendritic cells orchestrate the active protection of the lung tissue. During ageing the same is expected of the above cell types; however, immune cells change with age and their response to stimuli is no longer the same as in a young, healthy adult (see Chaps. 1 and 2).

The immunological defence of the lungs employs both innate and adaptive immune responses against antigens. Innate immunity is the critical first line of defence for the lungs while adaptive (acquired) immunity is antigen-specific and is required to ward off encapsulated bacteria, viruses, and intracellular pathogens. This form of immunity relies on immunological memory and antibody production. However, important changes in immunological responses occur with age and the impact on lung immunity can be summarised as follows:

6.4.1   Lung T-Cells

In histologically normal (not inflamed) human lung parenchyma there are approximately 1×10^{10} CD3 positive T-cells in residence. These T-cells are mainly tissue resident memory cells enriched for the immune response to local environmental antigens. This suggests that tissue resident T cells are present to maintain peripheral immune defence and that recruitment of memory T cells from blood or lymph nodes may not be necessary to recall an immune response in the lung. Memory T-cells originate from a pool of progenitor cells developing in the thymus that are undergoing a complex differentiation, selection and maturation process then encounter their T cell receptor (TCR)-specific antigen to finally mature into memory cells. However, naïve T-cell output progressively decreases with age due to involution of the thymus [28]. Epithelial cells that turn into adipocytes with age in the thymus, can no longer support T-cell differentiation and selection [28]. Consequently, the T-cell pool in the lung parenchyma can no longer be replenished by freshly released naïve T-cells reducing fast and effective immune reactions against novel antigens. Although lung-specific analysis of T-cell subtypes awaits further studies, data indicate that intrinsic deficiencies and defects in signalling, such as pathways involving T-cell cytokine production that lead to Th2 differentiation, are seen in old T cells [29].

6.4.2   Dendritic Cells (DCs)

Professional, antigen presenting DCs play a critical role in linking innate and adaptive immunity. Lung DCs are categorized as conventional DCs (cDCs), plasmacytoid DCs (pDCs) and monocyte-derived DCs (moDCs) each representing independent developmental lineages. Lung DCs develop in the bone marrow and enter the lung as pre-DCs that are thought to differentiate locally into mature DC subsets [30]. Although changes in pattern recognition and toll-like receptor expressions have not yet been reported during ageing in lung DCs, antigen uptake by pinocytosis has been shown to be less effective and cytokine expression profiles alter so that instead of Th1, they promote Th2-type T-cell responses [31]. Such a skewed response may help responses to extracellular infections but may dampen Th1 inflammatory responses to intracellular pathogens such as respiratory viruses.

6.4.3 Macrophages

Tissue macrophage subsets, including lung alveolar macrophages (AMs), often arise from embryonic progenitors that seed the organs and mature locally before and shortly after birth [31]. GM-CSF instructs lung foetal monocyte differentiation shortly before and after birth through activation of the nuclear receptor PPARγ [32]. Deletion of PPARγ in AMs of experimental animals resulted in pulmonary alveolar proteinosis (PAP) a process seen in patients with PAP who have low expression of PPARγ in AMs. AMs are maintained by proliferative self-renewal throughout life autonomously, independent from bone marrow–derived monocytes. It has been established that AMs suppress immune responses through the inhibition of DC-mediated activation of T cells and production of transforming growth factor β (TGFβ). Recent studies have also shown that TGFβ-induced Foxp3+ Treg cells (iTreg cells) inhibit spontaneous and antigen-induced development of Th2-type airway inflammation and induce tolerance to inhaled innocuous antigens [33]. The process is AM- and not DC-dependent and occurs in the lung tissue, not in the draining lymph nodes. In AMs, the switch from a tolerogenic mode to an inflammatory mode is accompanied by the secretion of IL-1, IL-6, TNFα and a latent TGFβ secretion where the latter only becomes activated by the integrin αVβ6 expressed on alveolar epithelial cells (AECs). In the absence of αVβ6 on epithelial cells spontaneous inflammation and emphysema develop [34]. Data indicate that detachment of AMs from epithelia upon infection may unleash AM inflammation by withdrawal of active TGFβ. Inhibition of pathological airway inflammation occurs via the intercommunication of AMs located in alveoli through the alveolar epithelium communication. The prevention of inflammatory responses is mediated by various inhibitory receptors on AMs, with the ligands expressed on AECs or present in the alveolar fluid [35].

With respect to ageing, recent studies resulted in controversial data when investigating senescent macrophages. In mouse studies some differences in TLR expression and cytokine responses as well as differentiation from macrophage progenitors were detected, but these have not all been observed or even investigated in human ageing studies. Nevertheless, aged macrophages exhibit low-grade pro-inflammatory phenotype and significantly reduced levels of autophagy that contribute to accelerated changes in the ageing process in general [36]. Unfortunately, little is known about specific changes characteristic to human alveolar or bronchial resident macrophages with age.

6.4.4 Lung Neutrophils

Neutrophils play a pivotal role in lung inflammation, but also clearance during and following inflammation. The role of neutrophils has been shown in a number of inflammatory lung diseases including ARDS, COPD, cystic fibroses, idiopathic pulmonary fibrosis, bronchiectasis and asthma [37]. It appears that following a successful inflammatory response neutrophils can change from a pro-inflammatory to an anti-inflammatory phenotype. In this case, neutrophils stop producing and releasing pro-inflammatory mediators (i.e. leukotriene B4, platelet-activating factor, IL-8) and instead begin to release resolving mediators including bioactive lipids (i.e. lipoxins, resolvins) that enhance resolution following inflammation [38].

With age there are dramatic changes in neutrophil function, including reduced chemotaxis, phagocytosis and bactericidal mechanisms (fully reviewed in Chap. 1). The reduced bactericidal function will predispose to infection but the reduced chemotaxis also has consequences for lung tissue as this results in increased tissue bystander damage from neutrophil elastases released during migration [39]. This reduced chemotactic behavior is due to dysregulated PI3 kinase intracellular signaling rather than reduced surface expression of chemoattractant receptors [40]. Neutrophil granulocytes are acknowledged as key players of COPD, increased neutrophil lung populations are associated with tissue damage, increased inflammation and impaired tissue repair. There is recent evidence suggesting that neutrophil functions (migration, ROS generation, degranulation, phagocytosis) are also all impaired in COPD resulting in bias towards increased inflammation and reduced bacterial clearance [40].

6.4.5 Epithelial Cells

The numerous roles of various pulmonary epithelial cells include mucin production, mucociliary clearance in conducting airways, reduction of surface tension in the alveoli of pulmonary host defences are well integrated with the ability of respiratory epithelial cells to respond to and 'instruct' the professional immune system to protect the lungs from infection and injury. For example surfactants lower surface tension are also involved in immune function. While the type II alveolar epithelial differentiation marker is surfactant protein C, the immune function of surfactants is primarily attributed to two surfactant proteins: A and D. They can both opsonize pathogens for phagocytosis. Low surfactant production, degradation or inactivation may therefore contribute to enhanced susceptibility to lung inflammation and infection. Furthermore, interaction of the signal-regulatory protein SIRPα (which mediates a so-called 'do not eat me' signal) on AMs with the globular heads of the surfactant proteins surfactant protein-A and surfactant protein-D suppresses AM inflammatory responses and phagocytosis, which can be overcome by TLR4 triggering that down-regulates SIRPα [41].

During ageing the pulmonary epithelium changes: the level of PPARγ decreases in both the alveolar epithelium and the supporting fibroblasts, consequently AECs become less capable of secreting surfactants which results in inefficient antimicrobial function and increased inflammatory cytokine production. Although it has not been reported in the context of pulmonary ageing, one can speculate that the process might be the same in several tissues where pro-fibrotic mediators are upregulated. There is evidence that the TGF-β/Smad3 pathway forms with αvβ6 integrin, mTOR and PPARγ a complex signalling network with extensive crosstalk regulating the development of fibrosis. In a rodent study it has been demonstrated that during colorectal fibrosis up-regulation of TGFβ, Smad3, αvβ6 and mTOR expression was detected while PPARγ expression was reduced [42].

6.5   PPARγ: A Prominent Member of the PPAR Family in the Ageing Lung

PPARγ expression and function appear to have a pivotal role in all molecular and cellular events associated with pulmonary senescence. PPARα, PPARβ, and PPARγ all show a common structure consisting of 4 domains: a variable amino terminal activation function-1 domain (AF-1), a DNA binding domain, a hinge region, and a conserved activation function-2 domain (AF-2). Domain AF-2 enables PPARs to bind structurally diverse natural and synthetic ligands [43]. In addition, AF-2 associates with co-regulators affecting receptor activity, receptor dimerization and nuclear translocation [44]. PPARs function as hetero-dimers with retinoid X receptors (RXR). Hetero-dimerization of PPAR with RXR is influenced by competing PPAR isoforms and other nuclear receptors. In their absence PPAR-RXR associates with co-repressor proteins of histone deacetylase activity. Ligands trigger co-repressor dissociation, and recruitment of co-activators [45]. Transcriptional activation or suppression may happen after recognition of PPAR response elements (PPRE) in target gene promoters and binding to PPRE consensus sequences [46].

PPARγ is a prominent member of the PPAR family and was first described as a regulator of adipocyte differentiation. Activation of PPARγ is triggered by a wide variety of natural as well as synthetic ligands. Natural PPARγ ligands include polyunsaturated fatty acids (PUFAs), eicosapentaenoic acids, and oxidized lipids [47]. The most studied family of synthetic ligands are the thiazolidinediones (TZDs). TZDs are used in the treatment of type 2 diabetes as they show insulin-sensitizing and hypoglycemic effects via activation of PPARγ [48]. Activation of PPARγ by TZDs results in the transcription of numerous genes involved in glucose and lipid utilization [49]. Examples for synthetic ligands include rosiglitazone (RGZ), ciglitazone (CGZ), pioglitazone (PGZ), and troglitazone (TGZ) [50].

6.5.1   PPAR Expression in Immune Cells and the Lung

PPARs are expressed in various cells of the immune system and the lungs (Table 6.1). PPARα and PPARγ are both expressed in macrophages and monocytes [51], eosinophils [52], with PPARβ also being expressed in neutrophils [37]. Dendritic cells only express PPARγ [53], while both PPARα and PPARγ are expressed by lymphocytes [54, 55]. PPARβ and PPARγ are both expressed in mast cells [56], while all three isoforms are present in airway epithelial cells [57]. As for mesenchymal cells, PPARγ is expressed by fibroblasts [58], while PPARα and PPARγ are present in airway smooth muscle cells [59]. These distinct patterns of expression suggest that activation of different isoforms may specifically regulate the production of inflammatory mediators and cellular responses.

There is increasing evidence suggesting that PPAR receptor patterns change in various lung disease, with PPARγ being the most extensively studied. Literature consensus for the role of PPARγ expression in the lungs is that it is up-regulated in response to diverse inflammatory conditions providing a negative feedback loop

Table 6.1 PPARγ-
expressing immune and
pulmonary cells

Cell type/subtype	PPARγ
Immune cells	
Lymphocytes	+
Monocyte/MØ	+
Neutrophils	+
Eosinophils	+
Dendritic cells	+
Pulmonary cells	
Epithelium	+
Fibroblasts	+
Smooth muscle	+

that allows natural PPARγ ligands to limit inflammatory responses in the lungs [60]. There is mounting evidence that PPAR ligands affect inflammatory processes through influencing cellular immune responses. These actions overlap with corticosteroids, exerting inhibitory effects on T cells, eosinophils, neutrophils, mast cells/basophils, and macrophages [61], with studies mostly focusing on PPARγ [62].

The PPARγ ligands PGJ2 and CGZ were reported to inhibit T cell proliferation [63]. PGJ2 efficiently induces T cell apoptosis and can also decrease the production of both Th1 and Th2 type cytokines from T cells [64]. Moreover, T cells treated with CGZ show decreased IFNγ, IL-4, and IL-2 secretion [65]. On the other hand, PGJ2 may also potentially trigger inflammation through the induction of IL-8 expression in T cells and macrophages via MAPK and NfkB signalling [66]. In monocytes, PGJ2 and TGZ efficiently inhibit the secretion of tumour necrosis factor α (TNFα), interleukin-1β (IL- 1β) and IL-6 [67]. PGJ2 and RGZ also decrease TNFα release and the expression of inducible nitric oxide synthase (iNOS) and matrix metalloproteinase (MMP)-9 in macrophages through inhibiting the activities of AP-1, STAT, and NFkB. Moreover, both PPARα and PPARγ ligands promote macrophage apoptosis as well [66, 67].

6.5.2 The Role of PPARγ in Pulmonary Tissue Homeostasis and Ageing

Tissue-specific stem cells have already been identified for many tissues. In the lungs alveolar type II cells (ATII), are essential for the development and repair of the gas-exchange surface. Surfactant protein production and survival of ATII cells is supported by lipofibroblasts, and their differentiation is PPARγ-dependent [68]. The process is strongly influenced by the Wnt/β-catenin signaling pathway: in the case of PPARγ-dominance lipofibroblast differentiation is skewed towards myofibroblast differentiation that does not support ATII replenishment. With age PPARγ expression decreases, while Wnt secretion increases. Consequently, stem cell capacity to renew the ATII cell pool decreases and as does pulmonary regenerative capacity shrink with age that renders the lungs more vulnerable to various diseases and conditions [6]. However, reinforcing PPARγ activity through TZD administration

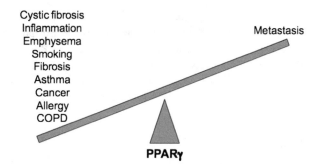

Fig. 6.3 Conditions related to diminished PPARγ activity (*left*) outweigh those linked with elevated PPARγ activity (*right*) in the lungs

has been shown to induce myofibroblast transdifferentiation into lipofibroblast cells. This may replenish the stem cell pool of pulmonary tissues, potentially counteracting the pro-ageing decrease of PPARγ activity observed during physiological senescence [69].

It is currently accepted that alterations in pulmonary PPAR profile, more precisely loss of PPARγ activity, can lead to inflammation, allergy, asthma, COPD, emphysema, fibrosis, and cancer (Fig. 6.3) [70]. Since it has been reported that PPARγ activity decreases with age, this provides a possible explanation for the increasing incidence of these lung diseases and conditions in older individuals [6]. The significance of maintaining PPARγ-activity in the lungs has led to research looking for potential novel biomarkers and therapeutic targets. A simplified approach would postulate TZDs as universal adjuvants for the treatment of various lung diseases and age-related pulmonary conditions. However, one must keep in mind that TZDs have a history in human therapy being used for the oral treatment of type 2 diabetes (PPARγ decreases insulin resistance) and certain TZDs have been restricted due to an increased incidence of cardiac events [71]. For pulmonary treatment this risk may be circumvented if nebulized TZDs are applied that maximize local efficiency and minimize systemic side effects [72]. Nevertheless, there are also reports on the potential harmful effects of TZDs with respect to lung cancer cell dissemination (not formation) by rendering the pulmonary micro-environment permissive for tumour cell survival [73].

6.6 The Role of PPARγ in Ageing Associated Pulmonary Diseases

Pulmonary tissue inflammation is often associated with ageing and may be triggered by various factors including infectious or chemical agents, but may also be initiated by endogenous factors as in the case of autoimmune conditions [74]. Despite the diversity of lung inflammation triggering factors the subsequently activated pro-inflammatory pathways are often shared. PPARγ can efficiently act on these pathways as it inhibits transcription factors regulating the expression of pro-inflammatory genes such as NF-kB, STAT-1 and AP1 [75, 76]. Activated PPARγ is capable of sequestering co-activator complexes on the promoter regions of pro-inflammatory genes that renders them inaccessible to these transcription factors. Also, PPARγ activation leads to the production of suppressive mediators, such as TGFβ and IL-10 that might be even more significant in the local micro-environment niche [77, 78]. As a consequence PPARγ up-regulation potentiates inhibitory loops that provide the molecular basis for resolution following an initially inflammatory response.

6.6.1   COPD

Chronic obstructive pulmonary disease (COPD) is common in older people, with an estimated prevalence of 10 % in the US population aged ≥75 years. Inhaled medications are the cornerstone of treatment for COPD and are typically administered by one of three types of devices (pressurized metered dose inhalers, dry powder inhalers, and nebulizers). However, age-related pulmonary changes may negatively influence the delivery of inhaled medications to the small airways [79]. In addition, physical and cognitive impairment, which are common in older patients with COPD, pose special challenges to the use of handheld inhalers in the old. Nebulizers should be considered for patients unable to use handheld inhalers properly. Airway mucus hypersecretion (AMH) is a key pathophysiological characteristic of COPD. Corticosteroid is the first-line anti-inflammatory treatment used to alleviate COPD, but its therapeutic effects are controversial and long term treatment often leads to undesirable side effects [80]. According to recent reports PPARγ agonists can inhibit mucin synthesis both in vitro and in vivo, but only nebulized TGZs lead to a reduction in mucus production in the airways, whereas oral administration has no such effect [81]. Administration of the PPARγ agonist ciglitazone via nebulizer reduces OVA-triggered mucus gland hyperplasia and airway occlusion by approximately 75 % [82]. It has been proposed that PPARγ activation may ease AMH through a pathway involving MMP-9, providing molecular mechanism of action for COPD treatment [83].

6.6.2   Lung Fibrosis

Currently approx. five million people are affected by pulmonary fibrosis worldwide. In most cases, patients are 40–50 years old at diagnosis, while the incidence of idiopathic pulmonary fibrosis increases drastically ≥50 years of age [84]. Fibrosis or fibrotic remodelling of lung tissue is a severe outcome of various lung diseases. Inflammatory cell invasion, epithelial cell injury and failure of re-epithelialization are followed by recruitment and persistence of fibroblasts. Excessive collagen and extracellular matrix production results in lung fibrosis [85]. Reduced PPAR expression was shown in lung fibroblasts of patients with dysregulated inflammation and fibrosis [86]. Studies found that TZDs were able to inhibit lung fibrosis [86]. PPAR ligands have negative effects on human lung fibroblasts by inhibiting proliferation and migration triggered by mitogenic growth factors such as PDGF [87]. Furthermore, PPAR agonists inhibit lung fibroblast differentiation mediated by

TGFβ and significantly reduce the expression of fibronectin and type I collagen [88]. Anti-fibrotic effects of TZDs has been demonstrated in animal models even after strong pro-fibrotic exposure (i.e. bleomycin) confirming them as potential candidates for therapy [89].

6.6.3   Lung Cancer

Lung cancers (LCs) have been defined as a major disease of the aged with disappointing survival statistics and represent the second most common type of cancer in both genders worldwide [90]. Another major inducer of LCs is cigarette smoking, although specific LC subtypes do not correlate with either age or smoking status (i.e. lung adenocarcinoma is more characteristic of young females) [91]. Cigarette smoke has also been shown to decrease PPARγ expression in the lungs leading to detrimental effects suggesting that PPARγ agonist treatment may counteract negative effects related to smoking and ageing. Indeed it was shown that treatment targeting PPARγ can significantly inhibit cigarette smoke-induced mucin production [92]. It has also been reported that exposure to acrolein, one of the most toxic compounds found in cigarette smoke, induces goblet cell hyperplasia in bronchial epithelium and induces airway inflammation, as shown by the increased levels of inflammatory cytokines including IL-1β, IL-8, and TNF-α in bronchial fluid. Treatment with TGZ before acrolein exposure was reported to alleviate these changes in a dose-dependent manner providing evidence for PPARγ efficiency in counteracting smoking [75]. These studies suggest a role of PPARγ in lung tumorigenesis, and suggest PPARs as potential biomarkers of lung tumours. Furthermore, the enhancement of PPARγ activity (through i.e. TZDs) could prevent the formation of lung cancer, or serve as adjuvant during lung cancer therapy. However, once tumour cells have been formed the situation may change. It has been reported that systemic administration of TGZ accelerates tumour metastasis in models of non-small cell lung cancer [93] and does not provide any survival benefit. Therefore systemic administration of PPARγ agonists should be avoided in advanced lung cancer, but again local administration perhaps via inhalers may reach the desired effect.

6.7   The Role of Wnt Pathways in Pulmonary Senescence: Regulation of PPARγ

Although PPARγ seems to play an important role in the differentiation and function of a great variety of pulmonary cell types, the question remains: why would PPARγ level and activity alter with age? Pinpointing the initial molecular trigger is difficult but recent studies highlighted the imbalance in Wnt signalling as age progresses [94]. Importantly, the opposition between PPARγ and the Wnt / β-catenin pathway has already been reported in several tissues by multiple research groups including ourselves [95]. It is conceivable that the pulmonary (epithelial) setting is similar in showing reciprocal changes of PPARγ and Wnt expression or activity with age. Indeed, increased Wnt4 and Wnt5a secretion in the ageing lung has been observed and both Wnts were reported to down-regulate PPARγ using a different mechanism [unpublished observations]. Reduction in PPARγ also leads to a decreased number of the alveolar progenitor or ATII cells; which can explain age-associated reduction in regenerative capacity. As ATII cells are also the source of surfactants, reduction in SP-levels indicates weakened functionality of ATII-s that was documented in both ageing animals as well as in the primary human lung tissue [96].

6.8   Conclusion

In summary, accumulated data suggest that although complexity of the ageing process is evident, by targeting some general molecular regulators (notably PPARγ signalling) using carefully designed, organ and tissue specific delivery methods, might allow us in the future to counteract the ageing program and delay decline of pulmonary function.

References

1. Weiss ST. Lung function and airway diseases. Nat Genet. 2010;42:14–6.
2. Boyle D. Is this the oldest person to have ever lived? Uzbekistan claims it has proof woman who died this week was 135—13 years older than French record holder. The Daily Mail. Accessed 11 April 2015.
3. Jeune B, Robine J-M, Young R, Desjardins B, Skytthe A, Vaupel JW. Jeanne Calment and her successors. Biographical notes on the longest living humans Bernard H. Maier et al. (eds.), Supercentenarians, Demographic Research Monographs, DOI 10.1007/978-3-642-11520-2_16, Springer-Verlag Berlin Heidelberg 2010.
4. Verbeken EK, Cauberghs M, Mertens I, Lauweryns JM, Van de Woestijne KP. Tissue and airway impedance of excised normal, senile, and emphysematous lungs. J Appl Physiol. 1992;72:2343–53.
5. Sharma G, Goodwin J. Effect of aging on respiratory system physiology and immunology. Clin Interv Aging. 2006;1(3):253–60.
6. Kovacs T, Csongei V, Feller D, Ernszt D, Smuk G, Sarosi V, et al. Alteration in the Wnt microenvironment directly regulates molecular events leading to pulmonary senescence. Aging Cell. 2014;13(5):838–49.
7. Tolep K, Higgins N, Muza S, Criner G, Kelsen SG. Comparison of diaphragm strength between healthy adult elderly and young men. Am J Respir Crit Care Med. 1995;152(2):677–82.
8. Polkey MI, Harris ML, Hughes PD, Hamnegärd CH, Lyons D, Green M, et al. The contractile properties of the elderly human diaphragm. Am J Respir Crit Care Med. 1997;155(5):1560–4.
9. Bailey KL, Bonasera SJ, Wilderdyke M, Hanisch BW, Pavlik JA, DeVasure J, et al. Aging causes a slowing in ciliary beat frequency, mediated by PKCε. Am J Physiol Lung Cell Mol Physiol. 2014;306(6):L584–9.
10. Ramos FL, Krahnke JS, Kim V. Clinical issues of mucus accumulation in COPD. Int J Chron Obstruct Pulmon Dis. 2014;9:139–50.
11. Hansel NN, Paré PD, Rafaels N, Sin DD, Sandford A, Daley D, et al. Genome-wide association study identification of novel loci associated with airway responsiveness in chronic obstructive pulmonary disease. Am J Respir Cell Mol Biol. 2015;53(2):226–34.
12. Demeo DL, Mariani TJ, Lange C, Srisuma S, Litonjua AA, Celedon JC, et al. The SERPINE2 gene is associated with chronic obstructive pulmonary disease. Am J Hum Genet. 2006;78(2):253–64.

13. Willis-Martineza D, Richards HW, Timchenkoa NA, Medranoa EE. Role of HDAC1 in senescence, aging, and cancer. Exp Gerontol. 2010;45(4):279–85.
14. Yao H, Rahman I. Role of histone deacetylase 2 in epigenetics and cellular senescence: implications in lung inflammaging and COPD. Am J Physiol Lung Cell Mol Physiol. 2012;303(7):L557–66.
15. Guarente L, Franklin H. Epstein lecture: sirtuins, aging, and medicine. New Engl J Med. 2011;364(23):2235–44.
16. Sasaki T, Maier B, Bartke A, Scrable H. Progressive loss of SIRT1 with cell cycle withdrawal. Aging Cell. 2006;5(5):413–22.
17. Barnes PJ. Cellular and molecular mechanisms of chronic obstructive pulmonary disease. Clin Chest Med. 2014;35(1):71–86.
18. Yao H, Rahman I. Current concepts on oxidative/carbonyl stress, inflammation and epigenetics in pathogenesis of chronic obstructive pulmonary disease. Toxicol Appl Pharmacol. 2011;54(2):72–85.
19. Menghini R, Casagrande V, Cardellini M, Martelli E, Terrinoni A, Amati F, et al. MicroRNA 217 modulates endothelial cell senescence via silent information regulator 1. Circulation. 2009;120(15):1524–32.
20. Lee J, Padhye A, Sharma A, Song G, Miao J, Mo YY, et al. A pathway involving farnesoid X receptor and small heterodimer partner positively regulates hepatic sirtuin 1 levels via microRNA-34a inhibition. J Biol Chem. 2010;285(17):12604–11.
21. Honeyman L, Bazett M, Tomko TG, Haston CK. MicroRNA profiling implicates the insulin-like growth factor pathway in bleomycin-induced pulmonary fibrosis in mice. Fibrogenesis Tissue Repair. 2013;6(1):16.
22. Strum JC, Johnson JH, Ward J, Xie H, Feild J, Hester A, et al. MicroRNA 132 regulates nutritional stress-induced chemokine production through repression of SirT1. Mol Endocrinol. 2009;23(11):1876–84.
23. Rane S, He M, Sayed D, Vashistha H, Malhotra A, Sadoshima J, et al. Downregulation of miR-199a derepresses hypoxia-inducible factor-1alpha and Sirtuin 1 and recapitulates hypoxia preconditioning in cardiac myocytes. Circ Res. 2009;104(7):879–86.
24. Hwa JJ, Suh Y. MicroRNA in aging: from discovery to biology. Curr Genomics. 2012;13(7):548–57.
25. Zhou R, O'Hara SP, Chen X-M. MicroRNA regulation of innate immune responses in epithelial cells. Cell Mol Immunol. 2011;8:371–9.
26. Panda A, Arjona A, Sapey E, Bai F, Fikrig E, Montgomery RR, et al. Human innate immunosenescence: causes and consequences for immunity in old age. Trends Immunol. 2009;30(7):325–33.
27. Whitsett JA, Alenghat T. Respiratory epithelial cells orchestrate pulmonary innate immunity. Nat Immunol. 2015;16(1):27–35.
28. Kvell K, Varecza Z, Bartis D, Hesse S, Parnell S, Anderson G, et al. Wnt4 and LAP2alpha as pacemakers of thymic epithelial senescence. PLoS One. 2010;5(5), e10701.
29. Busse PJ, Zhang TF, Srivastava K, Schofield B, Li XM. Effect of ageing on pulmonary inflammation, airway hyperresponsiveness and T and B cell responses in antigen-sensitized and -challenged mice. Clin Exp Allergy. 2007;37(9):1392–403.
30. Kopf M, Schneider C, Nobs SP. The development and function of lung resident macrophages and dendritic cells. Nat Immunol. 2015;16(1):1–9.
31. Agrawal A, Sridharan A, Prakash S, Agrawal H. Dendritic cells and aging: consequences for autoimmunity. Expert Rev Clin Immunol. 2012;8(1):73–80.
32. Moore KJ, Rosen ED, Fitzgerald ML, Randow F, Andersson LP, Altshuler D, et al. The role of PPAR gin macrophage differentiation and cholesterol uptake. Nat Med. 2001;T7:41–7.
33. Duan W, Croft M. Control of regulatory T cells and airway tolerance by lung macrophages and dendritic cells. Ann Am Thorac Soc. 2014;11 Suppl 5:S306–13.
34. Munger JS, Huang X, Kawakatsu H, Griffiths MJD, Dalton SL, Wu J, et al. A mechanism for regulating pulmonary inflammation and fibrosis: the integrin αvβ6 binds and activates latent TGF β1. Cell. 1999;96(3):319–28.

35. Kauffman HF. Innate immune responses to environmental allergens. Clin Rev Allergy Immunol. 2006;30(2):129–40.
36. Chuang SY, Lin CH, Fang JY. Natural compounds and aging: between autophagy and inflammasome. Biomed Res Int. 2014;2014: article ID:297293.
37. Robb CT, Regan KH, Dorward DA, Rossi AG. Key mechanisms governing resolution of lung inflammation. Semin Immunopathol. 2016. doi:10.1007/s00281-016-0560-6.
38. Serhan CN. Resolution phase of inflammation: novel endogenous anti-inflammatory and pro-resolving lipid mediators and pathways. Annu Rev Immunol. 2007;25:101–37.
39. Sapey E, Greenwood H, Walton G, Mann E, Love A, Aaronson N, et al. Phosphoinositide 3 kinase inhibition restores neutrophil accuracy in the elderly: towards targeted treatments for immunosenescence. Blood. 2014;123:239–48.
40. Sapey E, Stockley JA, Greenwood H, Ahmad A, Bayley D, Lord JM, et al. Behavioral and structural differences in migrating peripheral neutrophils from patients with chronic obstructive pulmonary disease. Am J Respir Crit Care Med. 2011;183:1176–86.
41. Sender V, Stamme C. Lung cell-specific modulation of LPS-induced TLR4 receptor and adaptor localization. Commun Integr Biol. 2014;7, e29053.
42. Latella G, Vetuschi A, Sferra R, Speca S, Gaudio E. Localization of $\alpha\nu\beta6$ integrin-TGF-β1/Smad3, mTOR and PPARγ in experimental colorectal fibrosis. Eur J Histochem. 2013;57(4), e40.
43. Nolte RT, Wisely GB, Westin S, Cobb JE, Lambert MH, Kurokawa R, et al. Ligand binding and co-activator assembly of the peroxisome proliferator-activated receptor-γ. Nature. 1998;395:137–43.
44. Berger J, Moller DE. The mechanisms of action of PPARs. Ann Rev Med. 2002;53:409–35.
45. Desvergne B, Wahli W. Peroxisome proliferator- activated receptors: nuclear control of metabolism. Endocr Rev. 1999;20:649–88.
46. Willson TM, Brown PJ, Sternbach DD, Henke BR. The PPARs: from orphan receptors to drug discovery. J Med Chem. 2000;43:527–50.
47. Bell-Parikh LC, Ide T, Lawson JA, McNamara P, Reilly M, FitzGerald GA. Biosynthesis of 15-deoxy-Δ12,14-PGJ2 and the ligation of PPARγ. J Clin Invest. 2003;112:945–55.
48. Benayoun L, Letuve S, Druilhe A, Boczkowski J, Dombret MC, Mechighel P, et al. Regulation of peroxisome proliferator-activated receptor γ expression in human asthmatic airways: relationship with proliferation, apoptosis, and airway remodelling. Am J Respir Crit Care Med. 2001;164:1487–94.
49. Faveeuw C, Fougeray S, Angeli V, Fontaine J, Chinetti G, Gosset P, et al. Peroxisome proliferator-activated receptor γ activators inhibit interleukin-12 production in murine dendritic cells. FEBS Lett. 2000;486:261–6.
50. Lehmann JM, Moore LB, Smith-Oliver TA, Wilkinson WO, Willson TM, Kliewer SA. An antidiabetic thiazolidinedione is a high affinity ligand for peroxisome proliferator-activated receptor γ (PPARγ). J Biol Chem. 1995;270:12953–6.
51. Standiford TJ, Keshamouni VG, Reddy RC. Peroxisome proliferator-activated receptor-γ as a regulator of lung inflammation and repair. Proc Am Thorac Soc. 2005;2:226–31.
52. Woerly G, Honda K, Loyens M, Papin JP, Auwerx J, Staels B, et al. Peroxisome proliferator-activated receptors α and γ down-regulate allergic inflammation and eosinophil activation. J Exp Med. 2003;198:411–21.
53. Gosset P, Charbonnier AS, Delerive P, Fontaine J, Staels B, Pestel J, et al. Peroxisome proliferator-activated receptor γ activators affect the maturation of human monocyte-derived dendritic cells. Eur J Immunol. 2001;31:2857–65.
54. Reynders V, Loitsch S, Steinhauer C, Wagner T, Stein- hilber D, Bargon J. Peroxisome proliferator-activated lymphocytes. Resp Res. 2006;7:104.
55. Jones DC, Ding X, Daynes RA. Nuclear receptor peroxisome proliferator-activated receptor α (PPARα) is ex- pressed in resting murine lymphocytes. The PPARα in T and B lymphocytes is both transactivation and transrepression competent. J Biol Chem. 2002;277:6838–45.
56. Sugiyama H, Nonaka T, Kishimoto T, Komoriya K, Tsuji K, Nakahata T. Peroxisome proliferator-activated receptors are expressed in human cultured mast cells: a possible role of these receptors in negative regulation of mast cell activation. Eur J Immunol. 2000;30: 3363–70.

57. Hetzel M, Walcher D, Grub M, Bach H, Hombach V, Marx N. Inhibition of MMP-9 expression by PPARγ activators in human bronchial epithelial cells. Thorax. 2003;58:778–83.
58. Burgess HA, Daugherty LE, Thatcher TH, Lakatos HF, Ray DM, Redonnet M, et al. PPARγ agonists inhibit TGF-β induced pulmonary myofibroblast differentiation and collagen production: implications for therapy of lung fibrosis. Am J Physiol. 2005;288:L1146–53.
59. Patel HJ, Belvisi MG, Bishop-Bailey D, Yacoub MH, Mitchell JA. Activation of peroxisome proliferator-activated receptors in human airway smooth muscle cells has a superior anti-inflammatory profile to corticosteroids: relevance for chronic obstructive pulmonary disease therapy. J Immunol. 2003;170:2663–9.
60. Ward JE, Tan X. Peroxisome proliferator activated receptor ligands as regulators of airway inflammation and remodelling in chronic lung disease. PPAR Res 2007: article ID: 14983.
61. Belvisi MG. Regulation of inflammatory cell function by corticosteroids. Proc Am Thoracic Soc. 2004;1:207–14.
62. Nie M, Corbett L, Knox AJ, Pang L. Differential regulation of chemokine expression by peroxisome proliferator- activated receptor γ agonists: interactions with glucocorticoids and β2-agonists. J Biol Chem. 2005;280:2550–61.
63. Cunard R, Ricote M, Di Campli D, Archer DC, Kahn DA, Glass CK, et al. Regulation of cytokine expression by ligands of peroxisome proliferator activated receptors. J Immunol. 2002;168:2795–802.
64. Harris SG, Phipps RP. ProstaglandinD2, its metabolite 15-d-PGJ2, and peroxisome proliferator activated receptor-γ agonists induce apoptosis in transformed, but not normal, human T lineage cells. Immunology. 2002;105:23–34.
65. Mueller C, Weaver V, Vanden Heuvel JP, August A, Cantorna MT. Peroxisome proliferator-activated receptor γ ligands attenuate immunological symptoms of experimental allergic asthma. Arch Biochem Biophys. 2003;418:186–96.
66. Fu Y, Luo N, Lopes-Virella MF. Upregulation of interleukin-8 expression by prostaglandin D2 metabolite 15- deoxy-Δ12,14 prostaglandin J2 (15d-PGJ2) in human THP-1 macrophages. Atherosclerosis. 2002;160:11–20.
67. Jiang C, Ting AT, Seed B. PPAR-γ agonists inhibit production of monocyte inflammatory cytokines. Nature. 1998;391:82–6.
68. Rehan VK, Sakurai R, Wang Y, Santos J, Huynh K, Torday JS. Reversal of nicotine-induced alveolar lipofibroblast to myofibroblast transdifferentiation by stimulants of parathyroid hormone-related protein signalling. Lung. 2007;185(3):151–9.
69. Wang Y. Peroxisome proliferator-activated receptor gamma agonists enhance lung maturation in a neonatal rat model. Pediatric Res. 2009;65(2):150–5.
70. Farrow SN. Nuclear receptors: doubling up in the lung. Curr Opin Pharmacol. 2008;8(3):275–9.
71. Cheng AY, Fantus IG. Thiazolidinedione-induced congestive heart failure. Ann Pharmacother. 2004;38(5):817–20.
72. Morales E, Sakurai R, Husain S, Paek D, Gong M, Ibe B, et al. Nebulized PPARγ agonists: a novel approach to augment neonatal lung maturation and injury repair in rats. Pediatr Res. 2014;75(5):631–40.
73. Lezzerini M, Budovskaya Y. A dual role of the Wnt signaling pathway during aging in Caenorhabditis elegans. Aging Cell. 2014;13:8–18.
74. Pollard KM. Silica, silicosis, and autoimmunity. Front Immunol. 2016;7:97.
75. Honda K, Marquillies P, Capron M, Dombrowicz D. Peroxisome proliferator-activated receptor γ is expressed in airways and inhibits features of airway remodeling in a mouse asthma model. J Allergy Clin Immunol. 2004;113:882–8.
76. Straus DS, Glass CK. Anti-inflammatory actions of PPAR ligands: New insights on cellular and molecular mechanism. Trends Immunol. 2007;28:551–8.
77. Freire-de-Lima CG, Xiao YQ, Gardai YQ, Bratton DL, Schiemann WP, Henson PM. Apoptotic cells, through transforming growth factor beta, coordinately induce anti-inflammatory and suppress pro-inflammatory eicosanoid and NO synthesis in murine macrophages. J Biol Chem. 2006;281:38376–84.

78. Yoon ZS, Kim SY, Kim MJ, Lim JH, Cho MS, Kang JL. PPARγ activation following apoptotic cell instillation promotes resolution of lung inflammation and fibrosis via regulation of efferocytosis and proresolving cytokines. Mucosal Immunol. 2015;8(5):1031–46.
79. Pritchard JN, Giles RD. Opportunities in respiratory drug delivery. Ther Deliv. 2014;5(12):1261–73.
80. Drummond MB, Dasenbrook EC, Pitz MW, Murphy DJ, Fan E. Inhaled corticosteroids in patients with stable chronic obstructive pulmonary disease: a systematic review and meta-analysis. JAMA. 2008;300:2407–16.
81. Mueller C, Weaver V, Vanden Heuvel JP, August A, Cantorna MT. Peroxisome proliferator-activated receptor γ ligands attenuate immunological symptoms of experimental allergic asthma. Arch Biochem Biophys. 2003;418:186–96.
82. Ward JE, Fernandes DJ, Taylor CC, Bonacci JV, Quan L, Stewart AG. The PPARgamma ligand, rosiglitazone, reduces airways hyperresponsiveness in a murine model of allergen-induced inflammation. Pulm Pharmacol Ther. 2006;19(1):39–46.
83. Shen Y, Chen L, Wang T, Wen F. PPARγ as a Potential target to treat airway mucus hypersecretion in chronic airway inflammatory diseases. PPAR Res. 2012:article ID:256874.
84. Sgalla G, Biffi A, Richeldi L. Idiopathic pulmonary fibrosis: diagnosis, epidemiology and natural history. Respirology. 2016;21(3):427–37.
85. Phan SH. The myofibroblast in pulmonary fibrosis. Chest. 2002;122(6):286–9.
86. Culver DA, Barna BP, Raychaudhuri B. Peroxisome proliferator-activated receptor gamma activity is deficient in alveolar macrophages in pulmonary sarcoidosis. Am J Respir Cell Mol Biol. 2004;30:1–5.
87. Aoki Y, Maeno T, Aoyagi K. Pioglitazone, a peroxisome proliferator-activated receptor gamma ligand, suppresses bleomycin-induced acute lung injury and fibrosis. Respiration. 2009;77:311–9.
88. Milam JE, Keshamouni VG, Phan SH. PPAR-gamma Agonists inhibit profibrotic phenotypes in human lung fibroblasts and bleomycin-induced pulmonary fibrosis. Am J Physio Lung Cell Mol Physiol. 2008;294:L891–901.
89. Dantas AT, Pereira MC, de Melo Rego MJ, da Rocha LF, Pitta RI, Marques CD, Duarte AL, Pitta MG. The Role of PPAR Gamma in Systemic Sclerosis. PPAR Res, 2015:article ID: 124624.
90. Jemal A, Siegel R, Ward E, Hao Y, Xu J, Thun MJ. Cancer statistics, CA-A. Cancer J Clin. 2009;59:225–49.
91. Zhai K, Ding J, Shi HZ. HPV and lung cancer risk: a meta-analysis. J Clin Virol. 2015;63:84–90.
92. Bein K, Leikauf GD. Acrolein—a pulmonary hazard. Mol Nutr Food Res. 2011;55(9):1342–60.
93. Li H, Sorenson AL, Poczobutt J, Amin J, Joyal T, Sullivan T, et al. Activation of PPARγ in myeloid cells promotes lung cancer progression and metastasis. PLoS One. 2011;6(12), e28133.
94. Florian MC, Nattamai KJ, Dorr K, Marka G, Uberle B, Vas V, et al. A canonical to non-canonical Wnt signalling switch in haematopoietic stem-cell ageing. Nature. 2013;503(21): 392–6.
95. Takada I, Kouzmenko AP, Kato S. Wnt and PPARgamma signaling in osteoblastogenesis and adipogenesis. Nat Rev Rheumatol. 2009;5(8):442–7.
96. Torday J, Rehan V. Neutral lipid trafficking regulates alveolar type II cell surfactant phospholipid and surfactant protein expression. Exp Lung Res. 2011;37(6):376–86.

Cancer, Ageing and Immunosenescence

7

Nora Manoukian Forones and Valquiria Bueno

Abstract

Worldwide the percentage of individuals older than 60 years is growing due to the increase in longevity. Age is an important risk factor for cancer and subjects aged over 60 also have a higher risk of comorbidities. Approximately 50 % of neoplasms occur in patients older than 70 years and these patients often have a range of therapies ranging from surgery, adjuvant or neoadjuvant therapy, or even palliative chemotherapy. While the OPAS (Pan American Health Organization) considers elderly subjects to be those aged over 60 years and for the WHO (World Health Organization) it is subjects over 65 years, a major concern for poor prognosis is with cancer patients over 70–75 years. These patients have a lower functional reserve, a higher risk of toxicity after chemotherapy, and an increased risk of infection and renal complications that lead to a poor quality of life. Several biomarkers (immunological and physiological) have been evaluated in cancer aiming to improve diagnostics, to predict whether a patient could have any benefit from a specific treatment, and to perform patient follow-up. Cancer is a complex disease and in this chapter we will discuss the distribution of cancer cases in older adults, the most important types of cancer in this population, treatment and outcomes, and possible immunosenescence characteristics related to cancer in the aged.

Keywords

Ageing • Cancer • Chemotherapy • Toxicity • Immunological biomarkers

N.M. Forones
Department of Medicine, UNIFESP Federal University of São Paulo, São Paulo, Brazil
e-mail: noraforones@gmail.com

V. Bueno (✉)
Department of Microbiology, Immunology and Parasitology, UNIFESP Federal University of São Paulo, Rua Botucatu 740, São Paulo, Brazil
e-mail: valquiriabueno@hotmail.com

List of Abbreviations

OPAS   Pan American Health Organization
WHO   World Health Organization
US   United States of America
UK   United Kingdom
ECOG   Eastern Cooperative Oncology Group
CGA   Comprehensive Geriatric Assessment
CRASH   Chemotherapy Risk Assessment Scale

7.1   Cancer in Older Adults

In 2005, individuals in the US aged ≥65 years contributed with 55 % of total cancers: 131,000 lung cancers (67 %), 90,000 colorectal cancers (64 %), 113,000 prostate cancer (61 %) and 79,000 female breast cancer (42 %) [1]. In the UK, cancer-incidence rate trends (1984–2007) were higher for prostate, lung, and colorectal cancer in males and for breast cancer in females aged ≥65 years [2]. In Japan, 60.8 % of prostate cancer patients were ≥75 years old and the average death in 2008 was 78.8 years [3]. According to GLOBOCAN 2012 Latin American and Caribbean estimated incidence per 100,000 increased with age for prostate (360.8/65–69 years, 529/70–74 years, 729/75+ years), colorectal (77.0/65–69 years, 109.2/70–74 years, 150.1/75+ years), lung (85.8/65–69 years, 113.7/70–74 years, 140.5/75+ years), and female breast cancer (176.3/65–69 years, 192.2/70–74 years, 210.5/75+ years). A similar pattern was observed for Australia/New Zealand: prostate (824.9/65–69 years, 941.4/70–74 years, 990.5/75+ years), colorectal (228.4/65–69 years, 304.2/70–74 years, 150.1/75+ years), lung (175.1/65–69 years, 240/70–74 years, 296/75+ years), and female breast (323.6/65–69 years, 322.3/70–74 years, 303.6/75+ years) cancer [4]. On the other hand, Egypt has a age-specific cancer incidence rate (2008–2011) for liver (314.8/65–69 years, 327.1/70–74 years, 363.5/75+ years) and bladder (128.6/65–69 years, 194.8/70–74 years, 205.6/75+ years) in males and breast (166.3/65–69 years, 138.7/70–74 years, 148.6/75+years) and liver (143.9/65–69 years, 167.9/70–74 years, 150.5/75+ years) in females [5]. China estimated incidence per 100,000 was higher with age for lung (178.4/65–69 years, 262.6/70–74 years, 551.2/75+ years), stomach (119.6/65–69 years, 162.1/70–74 years, 280.4/75+ years), and liver cancer (111.0/65–69 years, 139.6/70–74 years, 193.2/75+ years) [4]. It is noticeable that there is a difference in organs with higher cancer incidence in developed versus developing countries. Another observation is that cancer incidence increases with ageing almost irrespective of country and exceptionally for some organs there is a decrease in very old individuals. The findings from Surveillance, Epidemiology and End Results Program [6] show that almost a third of all cancer are diagnosed after the age of 75 years and 70 % of cancer-related deaths occur after the age of 65 years.

7.2   Clinical Trials in Older Patients

The traditional clinical trial focus is on younger and healthier patient, i.e. with few or no co-morbidities. These restrictions have resulted in a lack of data about the optimal treatment for older patients [7] and a poor evidence base for therapeutic decisions. Most physicians extrapolate data from the younger cancer patients which could lead to over treatment, although under treatment of older patients is also common due to their reduced ability to tolerate the full treatment regimen. The main reasons for this under representation in clinical trial include restrictions on eligibility. Standard performance status, as measured by Karnofsky performance or the Eastern Cooperative Oncology Group (ECOG), is inadequate to evaluate eligibility in older patients. Thus, components of clinical trials such as eligibility, endpoints, survivorship, and dose-limiting toxicity need to be re-evaluated in the context of the older patient. Other organ specific eligibility requirements may be more flexible, such as adequate renal function, when the drugs used are known not to be nephrotoxic. Barriers that prevent the participation of ageing patients in clinical trials include also social aspects: transportation; financial difficulties since most trials request multiple visits; limited expectation of remaining life time of the patient leads the family to avoid aggressive therapy and accept palliative approaches.

An International Society of Geriatric Oncology was founded in 2000 with the mission of developing health professionals in this area and to optimize the treatment of older patients [8, 9] as some age-specific instruments that could predict chemotherapy toxicity are needed [10] and the endpoint of the studies may also be different [11]. Most of them had as a main endpoint the patient survival. Due to the high risk of death in older patients due to other diseases, different endpoints need to be considered as response rate, clinical benefits, decreased pain scale, better quality of life may be more appropriate. Besides the common difficulties, older patients may also have cognitive differences that make it more difficult to understand the disease or personal care that should be taken during chemotherapy.

Another factor that should be investigated is the prior history of cancer. Patients aged 70 and more account for up to 50 % of the survivors of previous cancer [12–14]. These patients had specific problems, mainly associated with the remaining toxicities from the first treatment.

7.3   Multidisciplinary Treatment in Older Patients

The treatment decisions for older cancer patients should be multidisciplinary, with the purpose of evaluating the risk of treatment. Geriatricians use the comprehensive geriatric assessment (CGA) [15] to determine life expectancy and the risk of complications. The hand grip dynamometer allows the determination of the handgrip strength of the dominant hand and is a well validated measure of musculoskeletal function but also has a strong association with frailty and time to death. The timed up and go test involves asking the patient to get up from a chair without the use of arms, walk a distance of 3 m, turn around and return [16, 17]. Studies have shown that patients taking longer than 14 s had a high risk of falling and thus are frail or pre-frail and may be at high risk with aggressive treatment.

The CRASH Score (chemotherapy risk assessment scale) for older patients can also be used to predict the risk of chemotherapy toxicity in older patients. This score include a large number of variables such as predictors of hematologic (H) and non-hematologic (NH) toxicity. The predictors of grade 4 H toxicity were lymphocytes, aspartate aminotransferase level, Instrumental Activities of Daily Living score, lactate dehydrogenase level, diastolic blood pressure, and the toxicity of the chemotherapy regimen. Predictors of NH grade 3/4 were hemoglobin, creatinine clearance, albumin, Eastern Cooperative Oncology Group performance, Mini-Mental Status score, Mini-Nutritional Assessment score and the toxicity of the chemotherapy regimen [18].

Some common conditions in the older subjects such as malnutrition, anemia, neutropenia, and depression should also be treated before or alongside surgical or chemotherapy treatment of the cancer. Co-morbidities as diabetes, hypertension, emphysema, renal insufficiency should be improved after medical intervention and sometimes the dose adjustment of chemotherapy is required. The social support (transportation, nutrition and medical administration of drugs) may be evaluated as it influences the treatment.

The National Network of US Cancer [19] recommends that people over 70 should undergo some form of geriatric assessment. After appropriate analysis previously undiagnosed relevant conditions can be detected in more than 50 % of patients. There is clear evidence that CGA improves the functionality and quality of life and promotes the independence in older people undergoing cancer therapy, but its effects on survival are not yet well defined.

In the older patient, neutropenia, anemia, mucositis, cardiomyopathy and neuropathy—the toxic effects of chemotherapy—are more pronounced and may be related in part to physiological changes of age or of a higher prevalence of comorbidities [20–24]. The correction of comorbidities and malnutrition can lead to greater safety in the prescription of chemotherapy [25, 26].

7.3.1 Breast Cancer

Breast cancer appears frequently in females over 60 years. Older patients with early stage disease must be treated with surgical resection of the tumor [27]. Patients with low expectancy of life and hormone receptor positive tumor can be treated with only endocrine therapy: aromatase inhibitors offer a better control than tamoxifen to control the disease in these patients [28]. Sentinel node biopsy should be considered for patients in whom it might affect treatment decisions. Tumor stage I/II, in older women, are in 70 % HER2 negative and hormone receptor positive, these patients may be treated in advance with hormone therapy. Radiation therapy is not indicated for older patients. Studies show no difference in survival in older patients submitted to radiotherapy after resection.

7.3.2 Head and Neck Cancer

Surgical resection of the tumors can be done in patients over 70 years with the same benefits as younger patients. Toxicity is however higher in patients with comorbidities. Toxicity occurred in 63 % of older patients (>70 years) compared with 54 % of younger patients. Bilateral resection, male sex, being more than 70 years old and at an advanced stage were all associated with higher risk of toxicity [29]. Radiotherapy can also be indicated for older patients. Functional acute toxicity may be more severe in older patients (67 % versus 49 %), with no difference in survival [30]. Data from clinical trials using induced chemotherapy or chemoradiotherapy in patients of more than 70 years are rare.

7.3.3 Gastrointestinal Cancers

Tumors of the gastrointestinal tract commonly appear in subjects over 70 years old. The adjuvant treatment of colon cancer in patients of more than 70 years with fluoropyrimidines has the same benefits compared with the younger patients. The addition of oxaliplatin after age of 70 or 75 has to be evaluated according to the patient. Some studies show no benefit in the use of oxaliplatin as an adjuvant among older patients [31].

In metastatic patients, the chemotherapy with endovenous fluoropyrimidine is better tolerated than capecitabine. Oxaliplatin may be added to endovenous fluoro-pyridine. The risk of hematologic toxicities is higher among older colorectal cancer patients. Bevacizumab or cetuximab may be associated to the chemotherapy [32, 33]. Some studies have demonstrated a higher risk of thromboembolic disease in older patients using bevacizumab. There is currently not a consensus on the better treatment in rectal cancer in older patients. It seems that perioperative treatment, neoadjuvant radio chemotherapy and adjuvant chemotherapy may be indicated with a decreased risk of disease recurrence. In some studies this treatment was associated to a higher risk of death by other causes [34, 35].

The most prevalent primary cancer in the liver is hepatocellular carcinoma (HCC) and it is the fifth cause of cancer and the third cause of death by cancer in the world. Surgery, transplantation, TACE (transarterial chemoembolization), percutaneous ablation are the most common treatments. In patients over 70 years, HCC appears in hepatitis C or B patients, in non-alcoholic fatty liver disease or in subjects without any previous liver disease. These patients are commonly treated by palliative methods instead of surgical resection. However, a recent meta-anal-ysis did not show differences in survival after 3 years of treatment compared to younger patients [36].

Biliary tract cancers (BTC) include gallbladder cancer and intrahepatic, extrahe-patic or perihilar biliary duct cancer and are one of the less common cancer of the digestive system. The higher incidence of this cancer occur commonly in patients with more than 70 years and treatment of these cancers lack studies, probably due to the short incidence. BTC older patients were mostly associated with no surgery or chemotherapy treatment [37]. Studies compared surgery in stage I or II and showed no difference on survival. Chemotherapy was associated with an increased survival compared to best supportive care. Monotherapy with gemcitabine may be better supported than the association of gemcitabine to cisplatin.

7.3.4   Pancreatic Cancer

Pancreatic cancer presents as an aggressiveness disease, with the later diagnosis and the poor response to chemotherapy or molecular target drugs contributing to the poor prognosis of this cancer. The diagnosis of this disease is commonly made in patients over 70 years. The association of age and resection complication has been investigated and shown that the mortality increases during the first 30 and 90 days after surgery in older patients. However, among the patients that survived after this period the overall survival after 90 days is similar to the younger patients [38].

7.3.5   Lung Cancer

Non-small cell lung cancer: Older patients with NSCLC may tolerate surgical resection. Adjuvant therapy in patients with ECOG 0–1 has the same advantages for older compared to young patients. The same is observed in advanced disease. However, the addition of bevacizumab to the chemotherapy has not been associated to a better quality of life [39].

Small cell lung cancer: The treatment of this cancer with cisplatin and etoposide had the same response in older patients as that observed in younger patients. Carboplatin had the same effects as cisplatin with a decrease in renal toxicity but a higher hematological toxicity [19].

7.3.6   Melanoma

There is no difference between the surgery or radiotherapy treatment among patients over 65 years compared to those under 65 in localized melanoma. In metastatic melanoma the use of ipilimumab, BRAF inhibitors or dacarbazine also had the same results [19].

7.3.7   Urogenital Cancers

The main urogenital cancer is urothelial carcinoma. It appears frequently in the bladder after 65 years compared to the urethra, ureter or renal pelvis. The 5-year survival decreases progressively with the age being 84 % in patients with 65–69 years and 60 % in those with more than 85 years [40, 41].

For kidney cancer surgical resection is commonly indicated in the treatment of localized malignancy, but this surgery can be followed by renal insufficiency in patients over 75 years [42, 43]. Target treatment with sunitinib, sorafenib or temsirolimus, and everolimus in patients over 70 years has a higher risk of neutropenia, anemia and asthenia. Infections were also more common in older patients using everolimus. Interferon must not be used in older patients due to high toxicity [44–46].

Bladder cancer in 2014 was the 5th most incident (74,000 cases estimated) in USA and had the highest median age among all cancer types (73 years old). Muscle invasive bladder cancer (MIBC) is present in 25–30 % of all bladder cancer diagnoses and requires a more aggressive approach [47]. Older patients are often undertreated as in 40,000 patients with muscular invasive bladder cancer (MIBC), cystectomy varied between 55 % in those younger than 70 years of age to 21 % in those older than 79. In addition, in those non-receiving surgery patients (>79 years) only 30 % received chemotherapy and/or radiotherapy [48]. Recent adjusted analyses comparing older patients receiving radiation or cystectomy found no difference in overall survival or cancer-specific survival [49, 50]. Neoadjuvant chemotherapy based on methotrexate, vinblastine, doxorubicin, and cisplatin was shown by Grossman et al. to improve 5-year overall survival and became standard of care [51].

7.3.8   Prostate Cancer

The widespread use of prostate-specific antigen (PSA) screening has detected many early-stage prostate cancers (asymptomatic, clinically localized) increasing the probability of cure. Clinical (TNM) stage based on DRE, Gleason score in biopsy, and serum PSA best characterize prostate cancer. The choice for initial treatment is influenced by estimated life expectancy, comorbidities, therapy side effects and patient preference. Primary options for localized prostate cancer include active surveillance, radical prostatectomy and radiotherapy. Due to surgery risks, radical prostatectomy is used in general for patients whose life expectancy is greater than 10 years. For patients with widespread metastatic disease radiopharmaceuticals are an adequate option, particularly if they are not candidates for effective chemotherapy. This treatment is reported as potent for bone pain relief with minimal side effects [52].

Androgen deprivation therapy is the most common form of systemic therapy. However, it may lead to osteopenia and to an increased risk of bone fracture [53]. For patients with metastatic prostate cancer (M1) a survival benefit was obtained with bisphosphonates plus systemic chemotherapy. Weekly docetaxel for metastatic cancer was better tolerated, but had a decreased survival over the standard three-weekly regimen [54].

7.3.9   Ovarian Cancer

Older patients with ovarian cancer used to be treated with less aggressive treatment. The double treatment—surgery and chemotherapy—are not commonly used in this population as they result in poor survival compared to younger patients (18 % versus 53 %) [55, 56]. Peritoneal chemotherapy maybe indicated in some patients over 70 years but increased toxicity and reduced quality of life during treatment should be considered [57].

In conclusion, curative treatments cannot be denied to older patients based only on chronological age. The Comprehensive Geriatric Assessment and other forms of assessment must be used to determine the most appropriate treatment plan according to the functional status, social support and predicted life expectancy. Some older individuals may benefit from aggressive therapies, but they have a higher toxicity and may require a more complex clinical support. The estimated risk of severe toxicity with chemotherapy is only one aspect of treatment planning in the older patient and should be integrated into a multidisciplinary decision.

7.4   Immunosenescence and Cancer

The higher incidence of cancer observed with ageing can be explained by the accumulation of cell mutations during life, in addition to reduced potential for tissue regeneration and repair to cells and their constituents such as DNA. However, changes in the immune system also play an important role as the detection and elimination of tumor cells (immune surveillance, Fig. 7.1) could also be impaired in

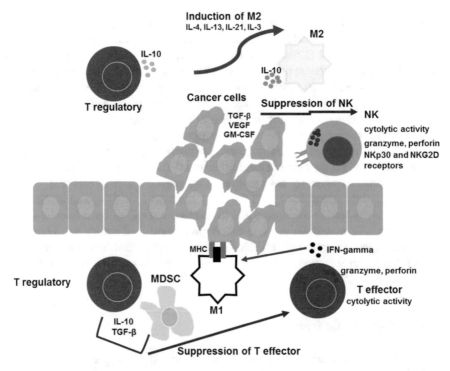

Fig. 7.1 Major effector and suppressive functions in the tumor microenvironment. Transformed (cancer cells) secrete factors that stimulate angiogenesis (VEGF) and increase macrophages and myeloid-derived suppressor cells (TGF-β, GM-CSG, M-CSF) at the tumor site. Macrophages (M1 subtype), T effector and NK cells infiltrate tumor and promote destruction by recognizing antigens in tumor cells via MHC (or its lack), secretion of factors such as reactive species of oxygen (ROS), reactive nitrogen species (RNS), IFN-γ, granzyme and perforin. However, tumor provides the induction of inflammatory/suppressive environment through factors such as IL-10, TGF-β, IL-4, IL-13, IL-21, and IL-3 that induces M2 macrophage, T regulatory (Treg) and myeloid-derived suppressor cells (MDSC). As consequence, the cytotoxic functions of T cells and NK cells are inhibited and there is a decrease in IFN-γ production and lower induction of MHC expression by macrophages. Killing-cell factors (granzyme and perforin) are diminished. M2 subset of macrophages secretes IL-10 and TGF-β and thus potentiate the suppressive tumor microenvironment

older individuals leading to increased susceptibility to cancer (Table 7.1). Together, these factors will determine how complex, aggressive and resistant to conventional treatment a specific cancer could be. Indeed, several authors have reported an inverse relationship between immune function and the incidence of cancer and in older adults as the immune function decreases (see Chaps. 1 and 2), the incidence of cancer increases [7, 8].

Table 7.1 Major age-related changes in immune system-related organs, tissues, and cells

Organ/Tissue	Changes	Consequences
Bone marrow	Hematopoietic progenitor generation is decreased Myeloid-biased hematopoietic stem cells	Decreased percentage of naïve T/B cells in periphery Increase of myeloid cells in periphery Increase of myeloid-derived suppressor cells in periphery causing a more suppressive environment
Thymus	Involution, increased adiposity, structural derangement of medulla and cortex	Impairment of thymopoiesis Lower frequency of naïve T cells in periphery, reduced TCR variability
Lymph nodes	Decrease of germinal centre follicles, increased adiposity, derangement of B and T cells area	Impairment in T follicular helper (Tfh) functions Decreased migration and function of dendritic cells
Spleen	Basic structure maintained	Decreased proportion of CD8+ T cells in the T cell zones, changes in CD4+ T cells both in T cell zones and in B cell areas
Cells	Changes	
Neutrophils/monocytes	Increased: numbers at periphery, danger signs recognition (inflammasome activation—IL-1 secretion) Decreased: capacity of migration, toll-like receptor (TLR1, 4) expression, phagocytosis, cytoplasmic antigen destruction, cytokines secretion after stimulation, NET (neutrophils extracellular traps) formation	
NK	Increased: numbers at periphery, CD56^{dim} phenotype Decreased: expression of inhibition receptors, release of perforin into the immunological synapse, cytotoxicity	
APCs macrophages, dendritic cells	Decreased: capacity of migration, antigen presentation	
T cells	Increased: memory phenotype, homeostatic proliferation, CD8+CD28-, production of Th2 cytokines, nTreg (natural T regulatory cells) Decreased: naïve phenotype, TCR repertoire, proliferation after stimulus, T helper function, production of Th1 cytokines	
B cells	Increased: production of antibodies (Ab) with low affinity, production of autoantibodies Decreased: production of Ab with high affinity and specificity	

Immunosenescence is considered a major contributor to cancer development in aged individuals and thus improved understanding of this issue is fundamental to develop new diagnostics methods and therapies for older patients, or to tailor existing therapies to take their immune functional decline in to account. Immunosenescence is a general classification for changes occurring in the immune system during the ageing process, as the distribution and function of cells involved in innate and adaptive immunity are impaired or remodelled [58, 59]. Some of these changes have been more extensively evaluated in clinical settings of cancer and will be discussed as they appear to play a role in cancer development.

7.4.1   Myeloid-Derived Suppressor Cells and Cancer

Myeloid-derived suppressor cells (MDSCs) represent a heterogeneous population of myeloid cells at different stages of differentiation comprising less than 5% of circulating cells in healthy adults. They originate from the same precursors in bone marrow that generate common myeloid progenitors such as monocytes, macrophages, dendritic cells and neutrophils. MDSCs percentages increase in cancer, infections, inflammatory conditions, and ageing [reviewed in 60]. In humans MDSCs have commonly be identified as lineage negative ($CD3^-$, $CD19^-$, $CD56^-$) and express $CD33^+$ HLA-DR$^{low/neg}$ $CD15^+$ (granulocytic MDSCs; g-MDSCs) and $CD33^+$ HLA-DR$^{low/neg}$ $CD14^+$ (monocytic MDSCs; m-MDSCs) [61, 62]. The generation of MDSCs is due to macrophage colony stimulating factor (M-CSF), granulocyte macrophage-colony stimulating factor (GM-CSF) and granulocyte-colony stimulating factor (G-CSF). Inflammatory conditions affect MDSC frequency leading to their expansion and recruitment by factors such as TGF-β, IL-1β, IL-6, IL-10, IL-12, IL-13, CCL-2, CXCL-5 and CXCL-12. In this context, tumors secrete factors that could act in the bone marrow to increase MDSCs and induce migration to the tumor site [reviewed in 60 and 63].

MDSCs have suppressive capabilities by producing reactive oxygen species ROS (highly expressed in g-MDSCs), arginase-1, and nitric oxide NO (mainly m-MDSCs). In addition, peroxynitrite (interaction of superoxide and NO) causes nitration of T cell receptor-CD8 complex, reducing this cell binding to the complex peptide — MHC I and thus rendering T cells unresponsive to antigen-specific stimulation. Depletion of L-arginine and cysteine caused my MDSCs leads to a decrease in CD3ζ chain expression, diminished IL-2 and IFN-γ production, and inhibition of T cell proliferation [64–67]. Other factors such as TGF-β, IL-10, and iNOS have also been reported as suppressive products of MDSCs [reviewed in 68].

Changes in the immune system during ageing occur in several tissues and in bone marrow, amongst others; it is observed a decreased lymphopoiesis and enhanced myelopoiesis [69, 70]. The skewing towards myelopoiesis with age in the bone marrow haematopoietic stem cells may contribute to the increase of myeloid-derived suppressor cells (MDSC) in older individuals [69] and potentiate tumor development.

Verschoor *et al.* [71] compared donors from community-dwelling and frail older adults with a history of cancer (breast, lung, prostate, skin, colon) and observed a significantly higher number of MDSC ($CD11b^+CD33^+HLA-DR^-$, $CD15^+CD11b^+$) in the latter group even though they were in partial or complete remission. In seven patients with advanced stage melanoma (<60 years old, n=2; 63–81 years old, n=5) treated with the CTLA-4 antagonist Ipilimumab, there was a decrease in MDSC after the first dose which remained low at week 9 in comparison with baseline levels [72]. Importantly, MDSC function was also reduced based on the production of arginase-1 which mediates many of the anti-proliferative effects of MDSCs. Iclozan et al. [73] evaluated patients (48–74 years old) with small-cell lung cancer who received a standard platinum/etoposide regimen and 4–6 weeks later were treated with dendritic cells (DC) or DC and all trans retinoic acid (ATRA) therapy which has been shown to reduce MDSC numbers. DC+ATRA decreased MDSC numbers, led to a higher number of patients with detectable p53 responses and increased the frequency of granzyme B^+ $CD8^+$ T cells. Finke et al. [74] have shown that favourable patient outcome is associated with low levels of intratumor/lesion MDSCs and levels of T cell responsiveness similar to healthy donors.

In addition, it has been reported that the presence of MDSC could favour the increase of T regulatory cells (Tregs) thus promoting a more suppressive environment and cancer development [75–77]. Authors believe that the presence of MDSC and/or Treg at the tumor site can subvert the effector function of T-infiltrating lymphocytes and generate a suppressive microenvironment that contributes to tumor development.

7.4.2   T Regulatory Cells and Cancer

In humans, most authors agree that numbers of naturally occurring T cells with regulatory function (Tregs) increase with age and that the frequency of inducible regulatory T cells declines [reviewed in 78]. The increase of Tregs in ageing patients could impose a regulatory/suppressive environment and thus contribute to tumor development.

El Andaloussi et al. [79] evaluated 10 patients (female mean age 60 years, male mean age 52 years) with glioblastoma and found in total $CD3^+$ T cells a significant increase in tumor and peripheral blood of $Foxp3^+$ T regulatory cells (55.1 % ± 1.88 % and 33.4 % ± 1.95 % respectively) in comparison with healthy donors (15.6 % ± 0.76 %). Sorted $CD4^+CD25^+$ (Treg) cells suppressed the proliferation of stimulated $CD4^+CD25^-$ T cells suggesting that the expansion of suppressive Treg cells in glioma patients may downregulate the antitumor response in the central nervous system. In agreement, Hiraoka et al. [80] found in patients with pancreas ductal adenocarcinoma (n=198 mean age 62±10 years) a prevalence of Treg cells in tissue-infiltrating (stroma) lymphocytes which was negatively correlated with patient survival, independent of tumor-node-metastasis classification, tumor grade, and tumor margins.

Chen et al. [81] observed an increase in Treg ($CD4+CD25+CD127^{low}$ and FoxP3 mRNA) from peripheral blood of aged healthy controls (60–74 years old) in comparison to young healthy controls (22–56 years old). In addition, Tregs frequency was significantly higher in aged urinary bladder cancer patients (UBC, 62–79 years old) than in aged healthy controls. These authors concluded that upregulated FoxP3 mRNA expression in association with increased Treg frequency could increase tumor escape, immune suppression and lead to ineffective tumor clearance in aged UBC patients.

7.4.3 T Effector T Cell Immune Responses in Cancer

Several authors have identified percentage, phenotype, and function of T cells in ageing cancer patients both pre and post treatment with the aim to use these immunologic markers as prognostic for patient outcome. Tumor-infiltrating lymphocytes seem to play an important role in anti-cancer immunity. Ebelt *et al.* [82] found in prostate cancer that T lymphocytes (mainly CD4+ T cells) form a cluster outside the carcinoma-infiltrating tissue in 17 non-treated patients (59–78 years old). However, almost no effector function was detected as IFN-gamma and perforin were downregulated on infiltrating lymphocytes compared to cells of healthy prostate tissue suggesting that infiltrating T cells with subverted function could be present at the tumor site and contribute for cancer development.

Ramirez *et al.* [83] evaluated patients with resected stage IIB-IV melanoma (15–87 years old) submitted to vaccination (12MP, multipotent vaccine derived from melanocytic differentiation antigens and cancer testis antigens) and verified the immune response (IR) based on ELIspot (interferon-gamma) from the peripheral blood mononuclear cells and increase on the CD8+ T cells. The 7 week cumulative incidence rate of IR was 53 % (130/249) in patients <64 years old and 38 % (30/78) in older patients over 64. Gender had no impact on IR and was not related with clinical outcome but there was a trend toward improved survival in males. The authors noted that even though patients older than 64 years old were able to mount an IR after vaccination, had 5 years disease free survival and overall survival not statistically different from patients under 64 years, their results may not apply to frail older adults as only patients with good functional status were included in the study.

Saavedra *et al.* [84] evaluated patients with stage IIIB or IV non-small cell lung cancer (NSCLC); 30 had not started first-line chemotherapy and 36 were treated with platinum-based chemotherapy. They received an EGF vaccine designed to induce specific antibodies against the epidermal growth factor (EGF) and had the percentage of T and B cells measured. CD4+, CD8+ and B cells were diminished in cancer patients regardless of chemotherapy in addition to the significant decrease in CD4/CD8 ratio. Terminally differentiated T cells (CD4+CD45RA+28− and CD8+CD45RA+CD28−) were not different when cancer patients and healthy individuals were compared. Vaccinated patients with high CD4+ T cells and CD4/CD8 ratio and low CD8+CD28− frequency as baseline values achieved a higher median survival suggesting that immunosenescence markers could provide support for determining whether patients could benefit from immunotherapy and be predictive of vaccine efficacy. In hepatocellular carcinoma patients (58–81 years old), a lower percentage of WTp53$_{264-272}$-specific naïve (CD45RA+) and higher percentages of WTp53$_{149-157}$ and WTp53$_{264-272}$-specific memory (CD45RO+) CD8+ T cells were observed and confirmed by CTL reactivity detection (granzyme B, IFN-gamma) suggesting interactions between the developing tumor and the host immune system [85].

7.4.4 Macrophages

Tumor cells use several mechanisms to invade extracellular matrix and metastasize to distant organs and the interaction between tumor cells and stromal cells in the tumor microenvironment plays an important role in tumor growth and metastasis [86–88]. Tumor-associated macrophages (TAMs) are prominent stromal cells in this interaction [89, 90].

Based on molecules expression, cytokines/chemokines and biogenic products, macrophages can be categorised as M1 which produce effector molecules (reactive oxygen intermediates, reactive nitrogen intermediates, TNFα) limiting tumor growth or M2 subtype. M2 macrophages promote tumor growth and metastasis by secretion of matrix-degrading enzymes, angiogenesis factors and immunosuppressive cytokines/chemokines [91–93]. The balance in macrophage subtypes will be determined by tumor microenvironment including other immune system-infiltrating cells and their products. In older humans, cancer and inflammatory chronic diseases have bring the majority of information about function and preferential polarisation to M1 or M2 in macrophages.

NSCLC patients without pre-operative chemotherapy/radiotherapy and short (average 1 year) or long survival (average 5.4 years) after resection (aged 58.0 ± 1.4 and 60.5 ± 1.3 respectively) had their lung tissue evaluated for the presence of macrophage subtypes M1 (CD68/HLA-DR) and M2 (CD68/CD163) in tumor islets, tumor stroma, or a combination of tumor islets and stroma [94]. M1 macrophage density in the tumor islets and stroma of patients with long survival time was significantly higher than the M1 macrophage density in the tumor islets and stroma of patients with short survival time. M2 macrophage density was not different when the lung tissue of short and long survival patients were compared. The authors also showed that M1 macrophage density in the tumor islets, stroma, or islets and stroma was positively associated with patient survival time whereas M2 density showed no association [94].

Patients with gastric cancer (≥ 70 years old) and peritoneal dissemination had a higher number of peritoneal macrophages with M2 phenotype (CD68/CD163 or CD68/CD204) than those without peritoneal dissemination. The authors concluded that the predominance of M2 phenotype could contribute to tumor proliferation/progression and is thus a promising target in the treatment of peritoneal dissemination in gastric cancer [95]. Gastric cancer patients (61.7 ± 1.2 years old) submitted to resection had their tissue specimens analysed for the presence of M1 and M2 macrophages. It was observed that patients with M1 density higher than 7.0 (n = 27) had a significantly superior survival rate (1-year survival rate = 81 %, 2-year survival rate = 55 %, 3-year survival rate = 26 %, 4-year survival rate = 22 % and 5-year survival rate = 18 %) compared to patients with lower M1 density. In the multivariate analysis M1/M2 ratio was a positive independent predictor of survival. M1 and M2 expression are potential markers to discriminate patients that could be more benefited from radically resection of gastric cancer [96]. Kurahara et al. evaluated 76 cases of pancreatic head cancer submitted to curative resection. Tumor infiltrating M2-polarized macrophages were associated with high incidence of lymph node metastasis and thus poor prognosis [97].

7.4.5   Natural Killer Cells

Natural killer (NK) cells comprise a heterogeneous population possessing immune surveillance functions to combat viral infections and malignancy based on direct cytotoxicity (granzyme/perforin and death receptor mediated) and/or secretion of cytokines and chemokines. In humans the NK phenotype CD3$^-$CD56$^+$ is further characterized into CD56^{dim} (90 %) and CD56^{bright} (10 %) subsets differing in receptor expression and function [98]. Le Garff-Tavernier et al. [99] showed that the frequency of CD3$^-$CD56$^+$ was significantly increased in very old subjects (87.1 ± 4.9 years) whereas the subset CD56^{bright} progressively decreased with age. The concomitant increase of CD56^{dim} NK suggested that the increase in NK frequency observed in older was related to CD56^{dim} maturation. This scenario could impact the ageing immune system as upon activation CD56^{dim} NK cells proliferate less, produce lower amounts of cytokines and are more cytotoxic whereas CD56^{bright} presents higher proliferation levels, produce cytokines (IFN-γ, TNF-β, IL-10) and chemokines (MIP-1a, RANTES) but is less cytotoxic [100].

NK cell cytotoxicity relies on the expression of molecules as CD16 (antibody-mediated toxicity), killer immunoglobulin-like receptors (KIR), LIR-1/ILT-2, CD94/NKG2A heterodimeric receptor which are inhibitory and recognize MHC I molecules as cognate ligands, and the collagen ligand LAIR-1 [101]. NK cells can promote cytolysis of cells lacking MHC I expression (i.e. cancer cells) but they will require an activating receptor such as NKG2C, NKG2D, NKp30, NKp44, NKp46, and NKp80 [102–105]. Interestingly Le Garff-Tavernier et al. [99] showed that the expression of LIR-1/ILT-2 was higher in older subjects compared to adults whereas the production of IFN-γ was impaired. Under IL-2 activation NK lysis was more evident in subjects older than 80 years but in absence of stimulus there was a decrease in this function.

Hazeldine et al. [106] also identified an increase in NK cell frequency with ageing in addition to a higher percentage of CD57$^+$ NK cells suggesting a shift towards a more mature circulating NK cell population. They also observed a significant decrease in NK cell cytotoxicity in older adults. The authors showed a decrease of NK cells expressing NKp30 or NKp46 and CD94 (binding partner of NKG2A). More importantly, they found that NK cells from older adults released less perforin into the immunological synapse, which was attributed to a defect in the polarisation of lytic granules to the NK target interface. The group stressed the importance of NK functions decline during ageing as this impairment in NK cells is related not only to lower function in detecting malignancy but has recently been associated with the reactivation of latent *Mycobacterium tuberculosis*, slower resolution of inflammatory responses, and increased incidence of bacterial and fungal infection [107].

Halama et al. [108] observed in patients with colorectal cancer (CRC; 40–78 years old) that the reduced or lost MHC I expression on tumor cells was not related to the recruitment of NK cells and the E-, L-, and P selectin that could interfere with migration of NK cells to tumor site were similar in normal mucosa and CRC. Contrary to the scarcity of NK cells in CRC, the levels of chemokines such as CXCL10, RANTES, CXCL9 were highly expressed along with high amounts of granzyme B and sCD95L which could be derived from infiltrating T cells instead. Papanikolaou et al. [109] found in CRC patients (mean age 75.9 years) a weak tissue presence of NK cells in 59.8 % of cases (n = 82) and a strong presence in 40.2 % of cases (primary tumors) whereas in metastatic lymph nodes there was 45.5 % of strong NK cell presence. It was observed a negative association between NK presence at the primary tumor site and the patients' age but this association was not seen at the metastatic lymph nodes.

Non small-cell lung cancer (NSCLC)-infiltrating NK cells were readily detected and displayed activation markers (NKp44, CD69, HLA-DR). However, the potential of cytolysis of NK from cancer tissues was lower than from peripheral blood or normal lung whereas their capability of producing cytokines was similar [110]. Platonova et al. [111] observed that NK were enriched in NSCLC site and localized in the stroma. Cells presented a decreased expression of NKp30, NKp80, DNAM-1, CD16, and ILT2 receptors compared with NK cells from distal lung tissues or blood. Moreover, the capabilities to stimulate degranulation and IFN-γ secretion were abolished in these cells in opposite to the circulating NK cells.

References

1. http://wonder.cdc.gov/CancerMort-v2005.html. Accessed 2 Sept 2010.
2. Ibrahim AS, Khaled HM, Mikhail NNH, Baraka H, Kamel H. Cancer incidence in Egypt; results of the national population-bases cancer registry program. J Cancer Epidemiol. 2014. doi:10.1155/2014/437971.
3. Mistry M, Parkin DM, Ahmad AS, Sasieni P. Cancer incidence in the United Kingdom: projections to the year 2030. Br J Cancer. 2011;105:1795–803.
4. Ministry of Health, Labour and Welfare: Patient Survey. Available from http://www.mhlw.go.jp/toukei/list/10-20.html. Accessed 10 Jul. 2015
5. GLOBOCAN 2012 Estimated Cancer Incidence. Mortality, and Prevalence Worldwide in 2012, Available from http://globocan.iarc.fr
6. http://seercancergov. Accessed on 2014.
7. Lichtman SM. Geriatric Oncology and Clinical Trials. ASCO Educational eBook; 2015 May 29-June2; Chicago-Illinois. American Society of Clinical Oncology, Alexandria VA.
8. Talarico L, Chen G, Pazdur R. Enrollment of elderly patients in clinical trials for cancer drug registration: a 7-year experience by the US Food and Drug Administration. J Clin Oncol. 2004;22:4626–31.
9. Wildiers H, Mauer M, Pallis A, Hurria A, Mohile SG, Luciani A, et al. Endpoints and trial design in geriatric oncology research: a joint European organisation for research and treatment of cancer-Alliance for Clinical Trials in Oncology International Society of Geriatric Oncology position article. J Clin Oncol. 2013;31:3711–8.
10. Hurria A, Togawa K, Mohile SG, Owusu C, Klepin HD, Gross CP, et al. Predicting chemotherapy toxicity in older adults with cancer: a prospective multicenter study. J Clin Oncol. 2011;29:3457–65.
11. Lichtman SM. Call for changes in clinical trial reporting of older patients with cancer. J Clin Oncol. 2012;30:893–4.
12. Cheung WY, Neville BA, Cameron DB, Cook EF, Earle C. Comparisons of patient and physician expectations for cancer survivorship care. J Clin Oncol. 2009;27:2489–95.
13. Earle CC. Cancer survivorship research and guidelines: may be the cart should be beside the horse. J Clin Oncol. 2007;25:3800–1.
14. De Santis CE, Lin CC, Mariotto AB, Siegel RL, Stein KD, Kramer JL, et al. Cancer treatment and survivorship statistics, 2014. CA Cancer J Clin. 2014;64:252–71.

15. Extermann M, Hurria A. Comprehensive geriatric assessment for older patients with cancer. J Clin Oncol. 2007;25(14):1824–31.
16. Rodin MB, Mohile SG. A practical approach to geriatric assessment in oncology. J Clin Oncol. 2007;25(14):1936–44.
17. Pondal M, Delser TJ. Normative data and determinants for the timed "up and go" test in a population-based sample of elderly individuals without gait disturbances. J Geriatr Phys Ther. 2008;31(2):57–63.
18. Extermann M, Boler I, Reich RR, Lyman GH, Brown RH, DeFelice J, et al. Predicting the risk of chemotherapy toxicity in older patients: the Chemotherapy Risk Assessment Scale for High-Age Patients (CRASH) score. Cancer. 2012;118(13):3377–86.
19. National Comprehensive Cancer Network. NCCN—Guidelines Index. Older adult oncology. 2015, Oncology. Available from http:www.nccn.org. Accessed on 21 Dec 2015.
20. Pasetto LM, Monfardini S. The role of capecitabine in the treatment of colorectal cancer in the elderly. Anticancer Res. 2006;26:2381–6.
21. Eichhorst BF, Busch R, Stilgenbauer S. First-Line therapy with fludarabine compared with chlorambucil does not result in a major benefit for elderly patients with advanced chronic lymphocytic leukemia. Blood. 2009;114:3382–91.
22. Biganzoli L, Licitra S, Moretti E, Pestrin M, Zafarana E, Di Leo A. Taxanes in the elderly: can we gain as much and be less toxic? Crit Rev Oncol Hematol. 2009;70:262–71.
23. Pal SK, Hurria A. Impact of age, sex, and comorbidity on cancer therapy and disease progression. J Clin Oncol. 2010;28:4086–93.
24. Pal SK, Katheria V, Hurria A. Evaluating the older patient with cancer: Understanding frailty and the geriatric assessment. CA Cancer J Clin. 2010;60:120–32.
25. Sak K. Chemotherapy and dietary phytochemical agents. Chemother Res Pract. 2012;2012:282570.
26. Balducci L. New paradigms for treating elderly patients with cancer: the comprehensive geriatric assessment and guidelines for supportive care. J Support Oncol. 2013;1(S2):30–7.
27. Morgan JL, Reed MW, Wyld L. Primary endocrine therapy as a treatment for older women with operable breast cancer—a comparison of randomised controlled trial and cohort study findings. Eur J Surg Oncol. 2014;40(6):676–84.
28. Sestak I, Baum M, Buzdar A, Howell A, Dowsett M, Forbes JF, et al. Effect of anastrozole and tamoxifen as adjuvant treatment for early-stage breast cancer: 10-year analysis of the ATAC trial. Lancet Oncol. 2010;11(12):1135–41.
29. Zabrodsky M, Calabrese L, Tosoni A, Ansarin M, Giugliano G, Bruschini R, et al. Major surgery in elderly head and neck cancer patients: immediate and long-term surgical results and complication rates. Surg Oncol. 2004;13(4):249–55.
30. Forastiere AA, Goepfert H, Maor M, Pajak TF, Weber R, Morrison W, et al. Concurrent chemotherapy and radiotherapy for organ preservation in advanced laryngeal cancer. N Engl J Med. 2003;349(22):2091–8.
31. Sanoff HK, Carpenter WR, Stürmer T, Goldberg RM, Martin CF, Fine JP, et al. Effect of adjuvant chemotherapy on survival of patients with stage III colon cancer diagnosed after age 75 years. Clin Oncol. 2012;30(21):2624–34.
32. Cassidy J, Saltz LB, Giantonio BJ, Kabinnavar FF, Hurwitz HI, Rohr UP. Effect of bevacizumab in older patients with metastatic colorectal cancer: pooled analysis of four randomized studies. J Cancer Res Clin Oncol. 2010;136(5):737–43.
33. Fornaro L, Baldi GG. Cetuximab plus irinotecan after irinotecan failure in elderly metastatic colorectal cancer patients: clinical outcome according to KRAS and BRAF mutational status. Crit Rev Oncol Hematol. 2011;8(3):243–51.
34. Lang K, Korn JR, Lee DW, Lines LM, Earle CC, Menzin J. Factors associated with improved survival among older colorectal cancer patients in the US: a population-based analysis. BMC Cancer. 2009;9:227.
35. Kim JH. Chemotherapy for colorectal cancer in the elderly. World J Gastroenterol. 2015;21:5158–66.
36. Hung AK, Guy J. Hepatocellular carcinoma in the elderly: meta-analysis and systematic literature review. World J Gastroenterol. 2015;21(42):12197–210.

37. Horgan A, Knox J, Aneja P, Le L, Mc Keever E, McNamara M, et al. Patterns of care and treatment outcomes in older patients with biliary tract cancer. Oncotarget. 2015;6(42): 44995–5004. doi:10.18632/oncotarget.5707.

38. Hayman TJ, Strom T, Springett GM, Balducci L, Hoffe SE, Meredith KL, et al. Outcomes of resected pancreatic cancer in patients age ≥70. J Gastrointest Oncol. 2015;6(5):498–504.

39. Maione P, Perrone F, Gallo C, Manzione L, Piantedosi F, Barbera S, et al. Pretreatment quality of life and functional status assessment significantly predict survival of elderly patients with advanced non-small-cell lung cancer receiving chemotherapy: a prognostic analysis of the multicenter Italian lung cancer in the elderly study. J Clin Oncol. 2005;23:6865–72.

40. Messing EM. Urothelial tumors of the bladder. In: Wein AJ, Kavoussi LR, Novick AC, Partin AW, Peters CA, editors. Campbell-Walsh urology. Ninthth ed. Philadelphia: Saunders-Elsevier; 2008. p. 2407–46. Chapter 75.

41. Schultzel M, Saltzstein SL, Downs TM, Shimasaki S, Sanders C, Sadler GR. Late age (85 years or older) peak incidence of bladder cancer. J Urol. 2008;179(4):1302–5.

42. Quivy A, Daste A, Haubaoui A, Duc S, Bernhard JC, Gross-Goupil M, et al. Optimal management of renal cell carcinoma in the elderly: a review. Clin Interv Aging. 2013;8:433–42.

43. Lane BR, Abouassaly R, Gao T, Weight CJ, Hernandez AV, Larson BT, et al. Active treatment of localized renal tumors may not impact overall survival in patients aged 75 years or older. Cancer. 2010;116(13):3119–26.

44. Schmidinger M, Larkin J, Ravaud A. Experience with sunitinib in the treatment of metastatic renal cell carcinoma. Ther Adv Urol. 2012;4:253–65.

45. Zanardi E, Verzoni E, Grassi P, Necchi A, Giannatempo P, Raggi D, et al. Clinical experience with temsirolimus in the treatment of advanced renal cell carcinoma. Ther Adv Urol. 2015;7:152–61.

46. Zustovich F, Novara G. Advanced kidney cancer: treating the elderly. Expert Rev Anticancer Ther. 2013;13:1389–98.

47. American Cancer Society. Cancer treatment and survivorship facts and figures 2014–2015. Atlanta: American Cancer Society; 2014.

48. Fedeli U, Fedewa SA, Ward EM. Treatment of muscle invasive bladder cancer: evidence from the National Cancer Database, 2003 to 2007. J Urol. 2011;185(1):72–8.

49. Smith AB, Deal AM, Woods ME, et al. Muscle-invasive bladder cancer: evaluating treatment and survival in the National Cancer Data Base. BJU Int. 2014;114(5):719–26.

50. Booth CM, Siemens DR, Li G, et al. Curative therapy for bladder cancer in routine clinical practice: a population-based outcomes study. Clin Oncol (R Coll Radiol). 2014;26(8): 506–14.

51. Grossman HB, Natale RB, Tangen CM, et al. Neoadjuvant chemotherapy plus cystectomy compared with cystectomy alone for locally advanced bladder cancer. N Engl J Med. 2003;349(9):859–66.

52. Mohler J, Bahnson RR, Boston R, Busby JE, D'Amico A, Eastham JA, et al. NCCN clinical practice guidelines in oncology: prostate cancer. J Natl Compr Canc Netw. 2010;8(2): 162–200.

53. Shahinian VB, Kuo YF, Freeman JL, Goodwin JS. Risk of fracture after androgen deprivation for prostate cancer. N Engl J Med. 2005;352(2):154–64.

54. Horgan AM, Seruga B, Pond GR, Alibhai SM, Amir E, De Wit R, et al. Tolerability and efficacy of docetaxel in older men with metastatic castrate-resistant prostate cancer (mCRPC) in the TAX 327 trial. J Geriatr Oncol. 2014;5(2):119–26.

55. Petignat P, Fioretta G, Verkooijen HM, Vlastos AT, Rapiti E, Bouchardy C, et al. Poorer survival of elderly patients with ovarian cancer: a population-based study. Surg Oncol. 2004;13(4):181–6.

56. Moore DH, Kauderer JT, Bell J, Curtin JP, Van Le L. An assessment of age and other factors influencing protocol versus alternative treatments for patients with epithelial ovarian cancer referred to member institutions: a Gynecologic Oncology Group study. Gynecol Oncol. 2004;94(2):368–74.

57. Armstrong DK, Bundy B, Wenzel L, Huang HQ, Baergen R, Lele S, et al. Intraperitoneal cisplatin and paclitaxel in ovarian cancer. N Engl J Med. 2006;354:34–43.
58. Fulop T, Kotb R, Fortin CF, Pawelec G, De Angelis F, Larbi A. Potential role of immunosenescence in cancer development. Ann NY Acad Sci. 2010;1197:158–65.
59. Hakim FT, Flomerfelt FA, Boyadzis M, Gress RE. Aging, immunity and cancer. Curr Opin Immunol. 2004;16:151–6.
60. Bueno V, Sant'Anna OA, Lord JM. Ageing and myeloid-derived suppressor cells: possible involvement in immunosenescence and age-related disease. Age. 2014;36:9729.
61. Zea AH, Rodriguez PC, Atkins MB, Hernandez C, Signoretti S, Zabaleta J, et al. Arginase-producing myeloid suppressor cells in renal cell carcinoma patients: A mechanism of tumor evasion. Cancer Res. 2005;65:3044–8.
62. Poschke I, Mougiakakos D, Hansson J, Masucci GV, Kiessling R. Immature immunosuppressive CD14+HLA-DR-/low cells in melanoma patients are Stat3hi and overexpress CD80, CD83, and DC-sign. Cancer Res. 2010;70:4335–45.
63. Ugel S, Delpozzo F, Desantis G, Papalini F, Simonato F, Sonda N, et al. Therapeutic targeting of myeloid-derived suppressor cells. Curr Opin Pharmacol. 2009;9:470–81.
64. Nagaraj S, Gupta K, Pisarev V, et al. Altered recognition of antigen is a mechanism of CD8+ T cell tolerance in cancer. Nat Med. 2007;13:828–35.
65. Lu T, Ramakrishnan R, Altiok S, et al. Tumor-infiltrating myeloid cells induce tumor cell resistance to cytotoxic T cells in mice. J Clin Invest. 2011;121:4015–29.
66. Rodriguez PC, Zea AH, Cullota KS, Zabaleta J, Ochoa JB, Ochoa AC. Regulation of T cell receptor CD3zeta chain expression by L-arginine. J Biol Chem. 2002;277:21123–9.
67. Srivastava MK, Sinha P, Clements VK, Rodriguez P, Ostrand-Rosenberg S. Myeloid-derived suppressor cells inhibit T-cell activation by depleting cystine and cysteine. Cancer Res. 2010;70(1):68–77.
68. Condamine T, Ramachandran I, Youn JI, Gabrilovich DI. Regulation of tumor metastasis by myeloid-derived suppressor cells. Annu Rev Med. 2015;66:97–110.
69. Beerman I, Bhattacharya D, Zandi S, Sigvardsson M, Weissman IL, Bryder D, et al. Functionally distinct hematopoietic stem cells modulate hematopoietic lineage potential during aging by a mechanism of clonal expansion. Proc Natl Acad Sci U S A. 2010;107(12):5465–70.
70. Chinn IK, Blackburn CC, Manley NR, Sempowski GD. Changes in primary lymphoid organs with aging. Semin Immunol. 2012;24(5):309–20.
71. Verschoor CP, Johnstone J, Millar J, Dorrington MG, Habibagahi M, Lelic A, et al. Blood CD33(+)HLA-DR(-) myeloid-derived suppressor cells are increased with age and a history of cancer. J Leukoc Biol. 2013;93:633–7.
72. Pico de Coaña Y, Pschkel I, Gentilcore G, Mao Y, Nyström M, Hanssom J, et al. Ipilimumab treatment results in an early decrease in the frequency of circulating granulocytic myeloid-derived suppressor cells as well as their Arginase1 production. Cancer Immunol Res. 2013;1:158–62.
73. Iclozan C, Antonia S, Chiappori A, Chen DT, Gabrilovich D. Therapeutic regulation of myeloid-derived suppressor cells and immune response to cancer vaccine in patients with extensive stage of small cell lung cancer. Cancer Immunol Immunother. 2013;62:909–18.
74. Finke J, Ko J, Rini B, Rayman P, Ireland J, Cohen P. MDSC as a mechanism of tumor escape from sunitinib mediated-anti-angiogenic therapy. Int Immunopharmacol. 2011;11:856–91.
75. Jordan KR, Amaria RN, Rairez O, Callihan EB, Gao D, Borakove M, et al. Myeloid-derived suppressor cells are associated with disease progression and decreased overall survival in advanced-stage melanoma patients. Cancer Immunol Immunother. 2013;62:1711–22.
76. Kalathil S, Lugade AA, Miller A, Iyer R, Thanavala Y. Higher frequencies of GARP(+) CTLA-4(+)Foxp3(+) T regulatory cells and myeloid-derived suppressor cells in hepatocellular carcinoma patients are associated with impaired T-cell functionality. Cancer Res. 2013;73:2435–44.
77. Yu J, Du W, Yan F, Wang Y, Li H, Cao S, et al. Myeloid-derived suppressor cells suppress anti-tumor immune responses through IDO expression and correlated with lymph node metastasis in patients with breast cancer. J Immunol. 2013;190:3783–97.

78. Jagger A, Shomojima Y, Goronzy JJ, Weyland CM. Regulatory T cells and the immune aging process: a mini review. Gerontology. 2014;60:130–7.
79. El Andaloussi A, Lesniak MS. An increase in CD4+CD25+Foxp3+ regulatory T cells in tumor-infiltrating lymphocytes of glioblastoma multiforme. Neuro Oncol. 2006;8:234–43.
80. Hiraoka N, Onozato K, Kosuge T, Hirohashi S. Prevalence of Foxp3+ regulatory T cells increases during the progression of pancreatic ductal adenocarcinoma and its premalignant lesions. Clin Cancer Res. 2006;12:5423–34.
81. Chen T, Wang H, Zhang Z, Li Q, Yan K, Tao Q, et al. A novel cellular senescence gene, SENEX, is involved in peripheral regulatory T cells accumulation in aged urinary bladder cancer. PLoS One. 2014;9(2), e87774.
82. Ebelt K, Barbara G, Figel AM, Phla H, Buchner A, Stief CG, et al. Dominance of CD4+ lymphocytic infiltrates with disturbed effector cell characteristics in the tumor microenvironment of prostate cancer. Prostate. 2008;68(1):1–10.
83. Ramirez AG, Wages NA, Hu Y, Smolkin ME, Slingluff Jr CL. Defining the effects of age and gender on immune response and outcomes to melanoma vaccination: a retrospective analysis of a single-institution clinical trials' experience. Cancer Immunol Immunother. 2015;64:1531–9.
84. Saavedra D, Garcia B, Lorenzo-Luaces P, González A, Popa X, Fuentes KP, et al. Biomarkers related to immunosenescence: relationships with therapy and survival in lung cancer patients. Cancer Immunol Immunother. 2015. doi:10.1007/s00262-015-1773-6.
85. Cicinnati VR, Zhang X, Yu Z, Ferencik S, Schmitz KJ, Dworacki G, et al. Increased frequencies of CD8+ T lymphocytes recognizing wild-type p53-derived epitopes in peripheral blood correlate with presence of epitope loss tumor variants in patients with hepatocellular carcinoma. Int J Cancer. 2006;119:2851–60.
86. Wittekind C, Neid M. Cancer invasion and metastasis. Oncology. 2005;69(S1):14–6.
87. Yoo BH, Wang Y, Erdogan M, Sasazuki T, Shirasawa S, Corcos L, et al. Oncogenic ras-induced down-regulation of pro-apoptotic protease caspase-2 is required for malignant transformation of intestinal epithelial cells. J Biol Chem. 2011;286:38894–903.
88. De Wever O, Mareel M. Role of tissue stroma in cancer cell invasion. J Pathol. 2003;200:429–47.
89. Pollard JW. Tumour-educated macrophages promote tumour progression and metastasis. Nat Rev Cancer. 2004;4:71–8.
90. Bingle L, Brown NJ, Lewis CE. The role of tumour-associated macrophages in tumour progression: implications for new anticancer therapies. J Pathol. 2002;196:254–65.
91. Stout RD, Jiang C, Matta B, Tietzel I, Watkins SK, Suttles J. Macrophages sequentially change their functional phenotype in response to changes in microenvironmental influences. J Immunol. 2005;175:342–9.
92. Murray PJ, Wynn TA. Obstacles and opportunities for understanding macrophage polarization. Eur J Immunol. 2007;37:S9–17.
93. Gordon S. The macrophage: past, present and future. J Leukoc Biol. 2011;89:557–63.
94. Ma J, Liu C, Che G, Yu N, Dai F, You Z. The M1 form of tumor-associated macrophages in non-small cell lung cancer is positively associated with survival time. BMC Cancer. 2010;10:112.
95. Yamaguchi T, Fushida S, Yamamoto Y, Tsukada T, Kinoshita J, Oyama K, et al. Tumor-associated macrophages of the M2 phenotype contribute to progression in gastric cancer with peritoneal dissemination. Gastric Cancer. 2015. doi:10.1007/s10120-015-0579-8.
96. Pantano F, Berti P, Guida FM, Perrone G, Vincenzi B, Amato MM, et al. The role of macrophages polarization in prediction prognosis of radically resected gastric cancer patients. J Cell Mol Med. 2013;17(11):1415–21.
97. Kurahara H, Shinchi H, Mataki Y, Maemura K, Noma H, Kubo F, et al. Significance of M2-polarised tumor-associated macrophage in pancreatic cancer. J Surg Res. 2011;167(2):e211–9.
98. Caligiuri MA. Human natural killer cells. Blood. 2008;112:461–9.
99. Le Garff-Tavernier M, Béziat V, Decocq J, Siguret V, Gandjbakhch F, Pautas E, et al. Human NK cells display major phenotypic and functional changes over the life span. Aging Cell. 2010;9(4):527–35.

100. Freud AG, Caligiuri MA. Human natural killer cell development. Immunol Rev. 2006;214:56–72.
101. Lebbink RJ, de Rulter T, Adelmeijer J, Brenkman AB, van Helvoot JM, Koch M, et al. Collagens arte functional, high affinity ligands for the inhibitory immune receptor LAIR-1. J Exp Med. 2006;203:1419–25.
102. Vivier E, Anfossi N. Inhibitory NK-cell receptors on T cells: wit-ness of the past, actors of the future. Nat Rev Immunol. 2004;4:190–8.
103. Bottino C, Castriconi R, Moretta L, Moretta A. Cellular ligands of activating NK receptors. Trends Immunol. 2005;26:221–6.
104. Lanier LL. NK cell recognition. Annu Rev Immunol. 2005;23:225–74.
105. Bryceson YT, March ME, Ljunggren HG, Long EO. Activation, coactivation, and costimulation of resting human natural killer cells. Immunol Rev. 2006;214:73–91.
106. Hazeldine J, Hampson P, Lord JM. Reduced release and binding of perforin at the immunological synapse underlies the age-related decline in natural killer cell cytotoxicity. Aging Cell. 2012;11:751–9.
107. Hazeldine J, Lord JM. The impact of ageing on natural killer cell function and potential consequences for health in older adults. Ageing Res Rev. 2013;12:1069–78.
108. Halama N, Braun M, Kahlert C, et al. Natural killer cells are scarce in colorectal carcinoma tissue despite high levels of chemokines and cytokines. Clin Cancer Res. 2011;17(4):678–89.
109. Papanikolaou IS, Lazaris AC, Apostolopoulos P, et al. Tissue detection of natural killer cells in colorectal adenocarcinoma. BMC Gastroenterol. 2004;4:20.
110. Carrega P, Morandi B, Costa R, et al. Natural killer cells infiltrating human nonsmall-cell lung cancer are enriched in CD56^{bright}CD16- cells and display an impaired capability to kill tumor cells. Cancer. 2008;112(4):863–75.
111. Platonova S, Cherfils-Vicini J, Damotte D, et al. Profound coordinated alterations of intratumoral NK cell phenotype and function in lung carcinoma. Cancer Res. 2011;71(16):5412–22.

Immune to Brain Communication in Health, Age and Disease: Implications for Understanding Age-Related Neurodegeneration

8

Jessica L. Teeling and Ayodeji A. Asuni

Abstract

The biggest threat to healthy ageing is the loss of our brain or eye function. Dementia and age-related vision loss are major causes of disability in our ageing population and it is estimated that a third of people aged over 75 are affected. Misfolded proteins, such as amyloid beta or alpha synuclein, pathological hallmarks of Alzheimer's disease and Parkinson's disease, are generally believed to be causative in the pathogenesis of these devastating disorders. However, analysis of post-mortem brain tissue from healthy older individuals has provided evidence that the presence of misfolded proteins alone does not correlate with cognitive decline and dementia, implying that additional factors are critical for neural dysfunction. We now know that innate immune genes and life-style contribute to the onset and progression of age-related neuronal dysfunction, suggesting that chronic activation of the immune system plays a key role in the underlying mechanisms that lead to irreversible tissue damage in the CNS. In this chapter we will discuss if, and how, the immune system regulates the CNS, which additional risk factor(s) contribute to the underlying mechanisms leading to neuronal dysfunction and whether intervention or immune modulation may be beneficial for those at risk of developing a devastating neurodegenerative disease. In particular, we will focus on the role of systemic infections and discuss the role of both the innate and the adaptive immune system in health, age and disease.

J.L. Teeling (✉)
Centre for Biological Sciences, University of Southampton, Southampton, UK
e-mail: J.Teeling@southampton.ac.uk

A.A. Asuni
Department Neurodegeneration In Vivo, H. Lundbeck A/S, Ottiliavej 9, 2500 Valby, Denmark

© Springer International Publishing Switzerland 2017
V. Bueno et al. (eds.), *The Ageing Immune System and Health*,
DOI 10.1007/978-3-319-43365-3_8

Keywords
Innate and adaptive immunity • Brain • Alzheimer's disease • Parkinson's disease • TNFα • Regulatory T cell • Systemic inflammation • Bacterial infection

8.1 Introduction

Diseases that cause progressive dysfunction of the brain and neuronal death are truly devastating as they are associated with memory loss, depression, and altered personalities. Most neurodegenerative diseases are associated with the accumulation and misfolding of proteins in specific regions of the CNS, for example amyloid beta in the cortex and hippocampus in Alzheimer's disease (AD) and alpha-synuclein in the substantia nigra in Parkinson's disease (PD) and it is generally believed that deposition of these misfolded proteins are causative in the pathogenesis. However, aggregates have also been detected in brain tissue from cognitively normal individuals [1], suggesting that other factors can drive neural dysfunction and/or loss. While age is the largest risk factor for the development of neurodegenerative diseases, in many cases the disease is influenced by environmental and genetic factors, and experimental, imaging and post-mortem studies of human brains have shown a correlation between microglial activation and behavioural, and cognitive changes prior to neuronal loss [2]. In addition, older patients with comorbidities such as atherosclerosis, type II diabetes or those suffering from repeated or chronic systemic bacterial and viral infections show earlier onset and progression of clinical symptoms [3–6] and recent GWAS studies have identified a number of susceptibility genes linked to both innate and adaptive immunity (e.g. TREM2, CD33, LRRK2, HLA-DR) [7–9]. Collectively these studies provide evidence for a critical role of inflammation in the pathogenesis of a range of neurodegenerative diseases, but the factors that drive or initiate inflammation remain largely elusive. In this chapter we will discuss if, and how, ageing affects immune regulation in the CNS, which additional risk factor(s) contribute to the underlying mechanisms leading to neuronal dysfunction or -loss and whether intervention or immune modulation may be beneficial for those at risk of developing a devastating neurodegenerative disease.

8.2 Immune to Brain Communication in Adulthood

It was once thought that the brain was isolated from peripheral immune responses, however it is now accepted that the brain plays an integral role in the body's response to inflammation, infection and/or stress. The effect of infection, mimicked experimentally by administration of bacterial lipopolysaccharide (LPS) has revealed that immune to brain communication is a critical component of a host organism's response to infection and a collection of behavioural and metabolic adaptations are initiated over the course of the infection with the purpose of restricting the spread of a pathogen, optimising conditions for a successful immune response and preventing the spread of infection to other organisms [10]. These behaviours are mediated by an innate immune response and have been termed '*sickness behaviours*' and include depression, reduced appetite, anhedonia, social withdrawal, reduced locomotor activity, hyperalgesia, reduced motivation, cognitive impairment and reduced memory encoding and recall [11, 12]. Metabolic adaptation to infection include fever, altered dietary intake and reduction in the bioavailability of nutrients that may facilitate the growth of a pathogen such as iron and zinc [10]. These behavioural and metabolic adaptions are evolutionary highly conserved and also occur in humans [13–15], further demonstrated by experiments where healthy volunteers are given a low dose of LPS and experience short term decline in cognition and increased depressive behaviours [16] and significant increased [11C]PBR28 binding, demonstrating microglial activation and neuroinflammation throughout the human brain following administration of LPS [17].

8.2.1 Underlying Mechanisms of Immune to Brain Communication

How the immune system initiates and maintains these adaptation depends on the locality, type and severity of the inflammatory stimulus and three principle routes of immune to brain communication have been described: (1) a neuronal route, via the vagus nerve; (2) a humoral route, via pro-inflammatory mediators in the circulation that interact with the intact cerebral vasculature and (3) a direct route via circumventricular organs (CVOs), which are structures in the brain that are characterized by their extensive vasculature, fenestrated capillaries and lack of a normal blood brain barrier (BBB) [18]. Evidence for neuronal signalling comes from studies on the vagus nerve, which expresses interleukin (IL)-1β, tumor necrosis factor α (TNFα) and prostaglandin receptors on the afferent fibers allowing detection of local inflammation [19–21]. While vagal signalling contributes to social withdrawal and fever in rodents, other behavioural adaptations such as anhedonia, anxiety, mood and memory do not only depend on neuronal signalling and involve a humoral route via proinflammatory mediators [22, 23]. These mediators that act on pathogen recognition receptors (PRRs) and cytokine receptors expressed on endothelial cells and perivascular macrophages at the BBB, resulting in *de novo* cytokine and PGE2 production by microglia [24–26]. The most widely studied cytokines produced in the CNS following peripheral LPS challenge are IL-1β and TNFα, which influence a number of behavioural changes and associated changes to microglia and neuronal function [22, 27]. The cell type responsible for the *de novo* transcription of cytokines remains unknown, although immunohistochemical analysis suggest that endothelial cells are the predominant cells in young, healthy rodent brains that produce IL-1β protein [28]. The behavioural and neuroinflammatory changes are not solely dependent on IL-1β and TNFα; IL-10 deficient mice show an increased and prolonged behavioural and neuroinflammatory response to systemic LPS administration, suggesting that regulation depends on a balance between pro- and anti-inflammatory mediators [29]. Furthermore, abrogation of peripheral cytokines

(IL-1β, IL-6 and TNFα) either separately or simultaneously with blocking antibodies has been shown to only partially attenuate behavioural changes and do not affect central cytokine production in response to LPS [22, 30]. This suggests that although they are not mutually exclusive; LPS can directly activate brain endothelial cells or induce other factors that promote *de novo* cytokine synthesis in the brain. These factors include the enzymes COX-1 and COX-2 that catalyse prostaglandin production [31, 32] and indoleamine 2,3-dioxygenase (IDO), an enzyme that degrades tryptophan and can be induced by INFγ and TNFα. Both enzymes are expressed in microglia and endothelial cells and pharmacological inhibition has been shown to abrogate depressive like behaviour in acute and chronic inflammation [33, 34]. Astrocytes may also contribute to immune to brain communication, either by producing pro-inflammatory mediators or by modulating the BBB, but these pathways and underlying mechanisms are beyond the scope of this chapter.

8.2.2 Immune to Brain Communication Following Real Live Infections

Whilst LPS is a useful tool for measuring the acute responses to an immune stimulus, it does not accurately model a real infection. Inflammatory reactions during infections usually have different kinetics and may engage a variety of pathogen recognition receptors and promote an adaptive immune response to control secondary infection. Currently, the number of publications on live infections and immune to brain communication is relatively small, when compared to microbial mimetics, but includes mycobacterium bovis strain Bacillus Calmette-Guerin (BCG), *Escherichia coli (E. coli)*, *Campylobacter jejuni*, *Bordetella pertussis*, *Salmonella typhimurium*, and influenza. BCG is a mycobacterium that is used as a vaccine against tuberculosis and stimulates an innate (macrophage) and adaptive (CD4 T cell) immune response, increasing circulating and local levels of TNFα and IFNγ [35]. Intraperitoneal injection of BCG causes an acute (<48 h) reduction in body weight, open field locomotor activity, and an elevation in core body temperature that last for 5 days [36]. Longer lasting effects include depressive behaviour that was observed up to 21 days after infection. The behavioural changes are associated with central TNFα and IFNγ production by microglia and endothelial cells [33, 37]. *E. coli* leads to acute levels of LPS in the circulation of adult rats, as well as significantly increased serum IL-1β, IL-6, and TNFα and central IL-1β levels beginning at 6-h post infection [38]. *E coli* infection leads to impaired memory function in 20 month-old, but not young rats, which may be mediated by the increased levels of IL-1β in the hippocampus and reduced levels of brain derived neurotrophic factor (BDNF) expression [39]. Studies with influenza have shown pronounced weight loss and accompanied anorexia, decreased sweetened milk intake and impaired reversal in the Morris water maze that measures long-term memory recall [30, 40]. These behavioural changes are accompanied by a two- to threefold increase in expression of IL-1β, IL-6 and TNFα in the hippocampus, measured 7 days after infection, and a decrease in expression of neurotrophins and immune regulatory molecules such as BDNF and CD200 [40]. Neutralization of IL-1β, IL-6 and TNFα only partially attenuated the effects of influenza on weight loss and sweetened milk intake, suggesting that this virus elicits immune to brain communication through an increased production of interferons and/or via the neuronal vagal route. It should be noted that reduction in body weight was severe (30 %), and the dose may be too high to translate to humans. Our own studies using a low dose of an attenuated strain of *Salmonella typhimurium* (SL3261) leads to a distinct neuroinflammatory and behavioural response compared to repeated injection of LPS [41]. Intraperitoneal infection with this intracellular bacterial strain results in robust macrophage and CD4 T cell activation in lymphatic tissue and circulating cytokines that peak 7 days post infection. These changes are associated with body weight loss (5–10 %), but sickness behaviours are only observed within 24 h after infection. In contrast to the systemic immune response, the brain responds with a delayed and prolonged cytokine production, which lasts up to 3 weeks, and coincides with increased CD4 T cell infiltration, and reduced activity in novel object recognition test, implying a decline in memory function (Hart et al., unpublished). Phenotype changes observed in microglia and endothelial cells include expression of MHCII and ICAM-1, suggesting a role of the BBB in these observations. These changes are observed up to 2 months post-infection, and are not observed following repeated LPS injection, clearly showing that a real life infection stimulates a more complex, but more physiological relevant immune to brain communication response.

8.3 Immune to Brain Communication in Ageing and Disease

Sickness behaviour and transient microglial activation are beneficial for individuals with a normal, healthy CNS, but in the ageing or diseased brain the response to peripheral infection can be detrimental and increases the rate of cognitive decline. Aged rodents exhibit exaggerated sickness and prolonged neuroinflammation in response to systemic infection, measured by increased levels of the proinflammatory cytokine IL-1β and microglia with increased levels of CD11b, CD68, MHCII, F4/80 [42–44]. Older people who contract a bacterial or viral infection or experience trauma postoperatively, also show exaggerated neuroinflammatory responses and are prone to develop delirium, a condition which results in a severe short term cognitive decline and a long term decline in brain function [45–47]. One explanation for these observations is that the innate immune cells of the CNS, the microglia are primed by their microenvironment due to neuronal changes and/or presence of misfolded proteins, and have a lower threshold for activation to secondary stimuli, including systemic infection. Priming is a well described response in macrophages *in vitro*, where an initial treatment of interferons or growth factors (e.g. IFNγ, GM-CSF, G-CSF) exacerbates their response to LPS [48]. Godbout's research group was the first to demonstrate the presence of primed microglia in aged rodents and showed twofold greater induction and prolonged expression of IL-1β and IL-6 mRNA in the brain after intraperitoneal injection of LPS in 20–24 month-old mice vs. 3–6 months adults. These molecular and cellular changes were accompanied with increased immobility in the forced swim test measured 72 h post infection and increased levels of indoleamine 2,3-dioxygenase (IDO), suggesting increased level of depression. Microglia are at least partly responsible for these effects as isolation of microglia from young and aged mice showed increased transcripts of IL-1β, iNOS and IDO [49–51], although other cell types, such as astrocytes and endothelial cells cannot be excluded. Peripheral surgical wounding in mice also activates microglia and increases the levels of TNFα and IL-6 in the hippocampus of both 9 and 18 month-old mice, but age potentiates these effects. In this study a critical role for peripheral macrophages and their production of TNFα was thought to drive cognitive impairment following aseptic trauma [52]. To unravel the underlying mechanisms, L'Episcopo et al., treated young and aged mice with HCT1026, a NO-donating derivative of flurbiprofen followed by a single sublethal dose of LPS. HCT1026 efficiently reversed the age-dependent increase of microglial activation in response to LPS to levels measured in younger mice and prevented the progressive loss of dopaminergic neurons in aged mice [53], suggesting that microglial priming occurs in multiple brain regions. Further experimental evidence for primed microglia, comes from studies using mice with ongoing neurodegeneration, which produce higher and prolonged levels of cytokines (IL-1β, TNFα) after local or peripheral LPS treatment resulting in nonreversible neuronal dysfunction and death [28]. Systemic administration of LPS also affects amyloid beta and tau pathology; two well described neuropathological hallmarks of AD. Repeated dosing of LPS can cause the accumulation of Aβ in both wild type and transgenic APP mice due to increased beta secretase activity [54, 55], and treating mice transgenic for human Tau (i.e. P301L, Tg4510) with LPS can result in the increased phosphorylation and accumulation of Tau in neurons [56]. Interestingly, a recently identified genetic risk factor for AD, CD33, was increased only in aged mice, which was correlated with increased Aβ beta load, cognitive impairment and ameliorated by treatment with ibuprofen, suggesting that systemic inflammation can modulate neuroinflammation and aggregation of misfolded proteins [57, 58]. Collectively these studies demonstrate that peripheral inflammation can increase the accumulation of two neuropathological hallmarks of AD, further strengthening the hypothesis that inflammation in involved in the underlying pathology.

8.3.1   Immune to Brain Communication Following Real Life Infections in Age and Disease

Despite epidemiological and clinical evidence for systemic infections as a driving force of neuronal dysfunction, mechanistic studies using real live microbial infection in experimental models of normal aging are still limited, and in particular chronic infection models. The studies that have been published to date show that old rats (24 month-old), experience prolonged loss of weight and core body temperature in response to live *E. coli*, associated with memory deficits [39]. A follow up study showed increased levels of proinflammatory cytokines in the liver and spleen, measured 4 days post infection, which was attenuated by depletion of hepatic and splenic macrophages, but the neuroinflammatory and memory deficits were not affected, suggesting that peripheral macrophages and release of pro-inflammatory cytokines do not play a role in the long-lasting adaptations in the hippocampus [59]. Others have reported that treatment of aged rats with dexamethasone treatment provides neuroprotection against *E. coli*-associated neurobehavioural and immuno-logical changes via anti-inflammatory and immunomodulatory effects [60]. Studies using live *E. coli* in normal aged mice have not yet been reported, but infection with an attenuated strain of Mycobacterium bovis, Bacillus Calmette-Guérin (BCG), shows prolonged sickness and depressive like behaviours in 20 month-old mice which can be detected up to 3 weeks post infection [61], but detailed analysis of microglia phenotype and function or immune modulation was not included in this study. Our own studies using a low dose (10^5 cfu) of *S. typhimurium*, showed exag-gerated and prolonged loss of weight, open field activity and static rod balance, and exaggerated IL-1β production in the hypothalamus, but not hippocampus, in the absence of robust phenotype changes when compared to non-infected aged mice (Hart et al., unpublished). Interestingly, infection of APP/PS1 with the life respira-tory pathogen *Bordetella* pertussis caused increased numbers of Aβ plaques and activation of microglia [62], but one of the less expected results was a robust infiltra-tion of CD4+ and CD8+ effector T cells [62]. Interestingly, recurrent infection of three common mouse models for AD, PD, and ALS (i.e., Tg2576, (Thy1)-[A30P] alpha SYN and Tg (SOD1-G93A)) with the most frequent respiratory pathogen, Streptococcus pneumonia, did not alter the course of neurodegeneration or levels of misfolded proteins, despite increased central production of IL-6 [63].

Studies from our own laboratory have shown that AD patients with mild cognitive impairment show a fivefold increased rate of cognitive decline when contracting a systemic urinary tract or respiratory tract infection, especially when serum levels of TNFα are elevated at baseline [64]. These observations are not restricted to normal aging and AD. Clinical manifestation of PD symptoms are increased following recent influenza infection (last infection 0–29 days: OR 3.03, 95 % CI 1.94–4.74), the num-ber of influenza episodes ≥3 attacks: OR 2.00, 95 % CI 1.45–2.75) and severity of the infection [65]. Apart from bacterial infection, chronic viral infections have also been linked to increased incidence of neurodegeneration, including cytomegalovirus (CMV). This virus is ubiquitously distributed in the human population, and along with other age-related diseases such as cardiovascular disease and cancer, has been associated with increased risk of developing vascular dementia and AD [66, 67]. The underlying mechanisms may include CMV induced immune changes, such as a reduction in naive T cells and/or increased number of T effector memory cells (CD4+/CD27−) and terminally differentiated memory T cells (CD27⁻CD45RA⁺), following chronic immune activation over a prolonged period [68].

8.3.2   Is There a Role for the Adaptive Immune System in Immune to Brain Communication?

Given the results of experimental models using real life infection, which, when given at physiological doses, show different outcomes when compared to selected innate immune stimuli, it is plausible that the adaptive immune system is involved in regulating how the immune system communicates with our brain. It is well known that adaptive immune responses progressively decline with age [69, 70].

Increasing experimental data suggest that T cells may also contribute to the regulation of microglia and neurons. For example, nude and severe combined immune deficient (SCID) mice that lack T cells, underperform in the Morris water maze, suggesting a reduced cognitive ability and adoptive transfer of splenocytes from wild-type donors resulted in improved learning behaviour [71, 72]. Similar results were obtained following the depletion of CD4+, but not CD8+ T cells [73], implying that CD4 T cells play a role in regulating neural function and/or plasticity. Modulation of T cells in experimental models of neurodegeneration further support these observations. Treatment with the immunosuppressive drug Tacrolimus (FK506), a molecule that inhibits the development and proliferation of T cells by neutralizing IL-2, significantly increased the survival of dopaminergic neurons following AAV-mediated overexpression of α-synuclein. Remarkably, no change in α-synuclein aggregation was observed, but FK506 significantly lowered the infiltration of both CD4+ T helper and CD8+ cytotoxic T cells and reduced the number of CD11b+microglial cells in the affected brain area, without changing CD68 and MHCII expression [74]. Interestingly, genetic loss of tau results in a reduction in microglia number and activation, and administration of FK506 to young Tau transgenic mice (P301S) attenuates tau pathology and increased lifespan, thereby linking neuroinflammation to early progression of tauopathies [75]. Similarly, acute treatment of APP transgenic mice (Tg2576) with FK506 improves memory function [76]. In accord, treatment of APP/PS1 mice with Copaxone (Glatiramer Acetate), a drug that breaks tolerance and promotes T cells development to brain antigens, results in decreased amyloid load associated with reduced cognitive decline and a microglial switch towards a neuroprotective phenotype [77]. However, aged APP/PS1 mice treated with live *Bordetella pertussis* results in a significant increase in effector CD4+ T cell infiltration that correlates with increase amyloid load [62]. These observations imply that the timing of activation and type of CD4+ T cell, may determine if the adaptive immune response is neuroprotective or not.

8.3.3 A Role for Regulatory T Cells?

With the aim of further understanding neuroinflammation in an experimental model of PD, Laurie et al., immunized mice with glatiramer and observed neuroprotection following 1-methyl-4-phenyl-1,2,3,6-tetrahydropyridine (MPTP) intoxication [78]. Adoptive transfer of lymphoid cells from glatiramer-immunized mice led to CD4+CD25+ T cell accumulation, increased IL-10, TGFβ and decreased TNFα, iNOS production in microglia, upregulation of glial cell derived neurotrophic factor, and significant rescue of Nissl+TH+ neurons in the SN [79]. Other studies have shown similar effects after immunization using Complete Freund's Adjuvant and, surprisingly, after infection with BCG [80]; the cell type responsible for the neuroprotective effect were identified as CD4+CD25+FoxP3+ T cells [81]. In addition, GM-CSF treatment reduces microgliosis, mediated by IL-27 production and promotion of natural Treg development, whilst increasing apoptosis of effector memory T cells, similar as reported for Rapamycin [82]. In the same MPTP model, adoptive transfer of natural regulatory T cells (Treg) specific for α-synuclein, are neuroprotective, while N-α-syn-specific helper T cells (Th1 and Th17) exacerbated neuronal degeneration [83]. Collectively these studies led to the conclusion that brain-specific Treg cell responses can home to damaged areas in the CNS and ameliorate neurodegenerative disease, but it should be noted that most studies used adult healthy mice treated with the neurotoxin MPTP. The use of aged mice has not yet been reported in the literature, but other models of neurodegeneration have. Synuclein vaccination of rats at early stages of PD (i.e., prior to neuronal dysfunction) induces a neuroprotective infiltration of CD4+/FoxP3+ Treg cell into the SN with sustained microglial activation, characterised by MHCII expression [84]. Neuroprotection was achieved by deleting α-synuclein specific effector T cells during thymic development in the mice; therefore, vaccination most likely induced a regulatory or tolerant immune response toward α-synuclein that conveys protection by preventing or stopping a detrimental immune response. To further confirm a protective effect of the adaptive immune system, Treg depletion in APP/PS1 mice, results in enhanced effector CD4+ T cell responses in AD compared with wild-type animals [85].

8.3.4 Evidence of Adaptive Immunity in Humans

The experiments using immunization and/or real infections imply that systemic infection can be both neuroprotective and exaggerate neuronal dysfunction, so how can we explain these opposing results? The answer may lie in timing of the infection and ability to regulate of the immune response, which is determined by both genetics and environmental risk factors. For example, healthy ageing is associated with a compromised adaptive immune response, but increased number and function of naturally occurring Treg cells [86]. In contrast, patients with symptomatic neurodegeneration lack this protection and instead have an 'exhausted' immune system. Evidence for this comes from immune phenotype studies: PD patients with more severe stages of disease (i.e. unified Parkinson's disease rating scale [UPDRS-III] of >30) have an increased circulating CD45RO+ and FAS+CD4+ T cells and decreased α4β7+ and CD31+ CD4+ T cells, indicative of increased effector/memory T cells likely due to chronic immune activation in disease progression. Furthermore, these cells do not suppress CD4 and CD8 cells *in vitro*—and do not proliferate after a CD3/CD28 stimulation [87, 88]. Similarly, immunophenotyping AD patients revealed a lower proportion of naïve T cells, more late-differentiated cells and higher percentages of activated CD4+CD25+ T cells without a Treg phenotype, implying that the neuroprotective effect of nTreg cells is lacking in older people with a neurodegenerative disease [89, 90]. These changes to CD4+ cells maybe the result of persistent antigenic challenge, perhaps induced by chronic immune activation and microbial infections. CD4+CD25(high) potentially Treg cells with a naive phenotype are also reduced in AD patients.

In conclusion, inflammation is increasingly linked to the pathogenesis of age-related neurodegenerative diseases. In this chapter we have described how systemic infection may drive neuroinflammation and innate immune activation in the brain, especially during ageing. The underlying mechanism are complex but experimental and clinical studies suggest that dysregulation of both our innate and adaptive immune system are responsible for the exaggerated responses observed following systemic infection. Figure 8.1 summarises the potential immune signalling scenarios in the young, aged and diseased brain. A recent clinical trial using etanercept, a selective TNFα inhibitor shows promising effect in delaying cognitive decline [91],

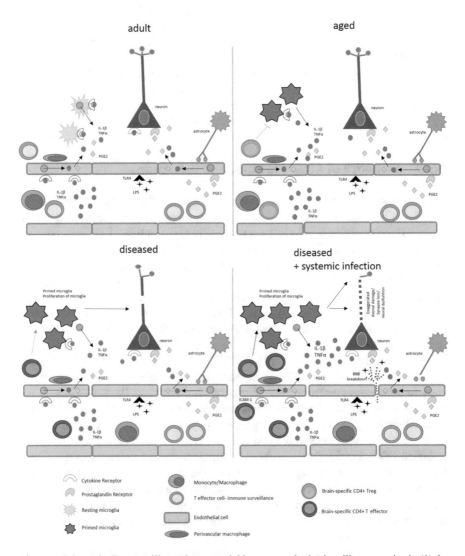

Fig. 8.1 Schematic diagrams illustrating potential immune-to-brain signalling scenarios in (1) the normal young adult brain, (2) the normal aged adult brain, (3) the pathological disease adult brain and (4) the pathological disease adult/aged brain and systemic infection

but further work is needed to test if immune modulation of the adaptive immune system to boost the function of regulatory T cells is beneficial for those at risk of developing a devastating neurodegenerative disease.

Systemic inflammation results in increased levels of cytokines and PGE2 in the adult young brain, and transient de novo inflammatory mediators production by cerebral endothelial cells, perivascular macrophages and microglia. These events lead to sickness behaviours, such as anhedonia, but not neuronal damage. Systemic inflammation results in exaggerated cytokine production in the aged brain, due to the presence of primed microglia, but the infiltration of brain-specific Treg cells can regulate neuroinflammation and neuronal damage. In the older diseased brain, systemic inflammation results in exaggerated cytokine production by primed microglia, triggered by brain specific T effector cells which contributes to neuronal damage, and thus cognitive decline; these events are exaggerated following systemic inflammation. This imbalance in adaptive immune regulation maybe the result of life long exposure to environmental stimulation causing exhaustion of the aged immune system and promotion of low grade inflammation. These adaptation may be one explanation for why systemic infections contribute to the onset and progression of age-related neurodegenerative diseases.

References

1. Elman JA, Oh H, Madison CM, Baker SL, Vogel JW, Marks SM, et al. Neural compensation in older people with brain amyloid-beta deposition. Nat Neurosci. 2014;17(10):1316–8.
2. Meyer-Luehmann M, Prinz M. Myeloid cells in Alzheimer's disease: culprits, victims or innocent bystanders? Trends Neurosci. 2015;38(10):659–68.
3. Habeych ME, Castilla-Puentes R. Comorbid medical conditions in vascular dementia: a matched case-control study. J Nerv Ment Dis. 2015;203(8):604–8.
4. Sandu RE, Buga AM, Uzoni A, Petcu EB, Popa-Wagner A. Neuroinflammation and comorbidities are frequently ignored factors in CNS pathology. Neural Regen Res. 2015;10(9):1349–55.
5. Perry VH, Cunningham C, Holmes C. Systemic infections and inflammation affect chronic neurodegeneration. Nat Rev Immunol. 2007;7(2):161–7.
6. Rocha NP, de Miranda AS, Teixeira AL. Insights into neuroinflammation in Parkinson's disease: from biomarkers to anti-inflammatory based therapies. Biomed Res Int. 2015;2015:628192.
7. Hollingworth P, Harold D, Sims R, Gerrish A, Lambert JC, Carrasquillo MM, et al. Common variants at ABCA7, MS4A6A/MS4A4E, EPHA1, CD33 and CD2AP are associated with Alzheimer's disease. Nat Genet. 2011;43(5):429–35.
8. Holmans P, Moskvina V, Jones L, Sharma M, International Parkinson's Disease Genomics C, Vedernikov A, et al. A pathway-based analysis provides additional support for an immune-related genetic susceptibility to Parkinson's disease. Hum Mol Genet. 2013;22(5):1039–49.
9. International Parkinson Disease Genomics C, Nalls MA, Plagnol V, Hernandez DG, Sharma M, Sheerin UM, et al. Imputation of sequence variants for identification of genetic risks for Parkinson's disease: a meta-analysis of genome-wide association studies. Lancet. 2011; 377(9766):641–9.
10. Hart BL. Biological basis of the behavior of sick animals. Neurosci Biobehav Rev. 1988;12(2):123–37.
11. Dantzer R. Cytokine-induced sickness behaviour: a neuroimmune response to activation of innate immunity. Eur J Pharmacol. 2004;500(1-3):399–411.
12. Dantzer R. Innate immunity at the forefront of psychoneuroimmunology. Brain Behav Immun. 2004;18(1):1–6.

13. Bucks RS, Gidron Y, Harris P, Teeling J, Wesnes KA, Perry VH. Selective effects of upper respiratory tract infection on cognition, mood and emotion processing: a prospective study. Brain Behav Immun. 2008;22(3):399–407.
14. Smith AP. Effects of the common cold on mood, psychomotor performance, the encoding of new information, speed of working memory and semantic processing. Brain Behav Immun. 2012;26(7):1072–6.
15. Smith AP. Effects of upper respiratory tract illnesses and stress on alertness and reaction time. Psychoneuroendocrinology. 2013;38(10):2003–9.
16. Krabbe KS, Reichenberg A, Yirmiya R, Smed A, Pedersen BK, Bruunsgaard H. Low-dose endotoxemia and human neuropsychological functions. Brain Behav Immun. 2005;19(5):453–60.
17. Sandiego CM, Gallezot JD, Pittman B, Nabulsi N, Lim K, Lin SF, et al. Imaging robust microglial activation after lipopolysaccharide administration in humans with PET. Proc Natl Acad Sci U S A. 2015;112(40):12468–73.
18. Fry M, Ferguson AV. The sensory circumventricular organs: brain targets for circulating signals controlling ingestive behavior. Physiol Behav. 2007;91(4):413–23.
19. Ek M, Kurosawa M, Lundeberg T, Ericsson A. Activation of vagal afferents after intravenous injection of interleukin-1beta: role of endogenous prostaglandins. J Neurosci. 1998;18(22):9471–9.
20. Goehler LE, Gaykema RP, Hammack SE, Maier SF, Watkins LR. Interleukin-1 induces c-Fos immunoreactivity in primary afferent neurons of the vagus nerve. Brain Res. 1998;804(2):306–10.
21. Rogers RC, Hermann GE. Tumor necrosis factor activation of vagal afferent terminal calcium is blocked by cannabinoids. J Neurosci. 2012;32(15):5237–41.
22. Teeling JL, Felton LM, Deacon RM, Cunningham C, Rawlins JN, Perry VH. Sub-pyrogenic systemic inflammation impacts on brain and behavior, independent of cytokines. Brain Behav Immun. 2007;21(6):836–50.
23. Wieczorek M, Dunn AJ. Relationships among the behavioral, noradrenergic, and pituitary-adrenal responses to interleukin-1 and the effects of indomethacin. Brain Behav Immun. 2006;20(5):477–87.
24. Zhang J, Rivest S. A functional analysis of EP4 receptor-expressing neurons in mediating the action of prostaglandin E2 within specific nuclei of the brain in response to circulating interleukin-1beta. J Neurochem. 2000;74(5):2134–45.
25. Rivest S. How circulating cytokines trigger the neural circuits that control the hypothalamic-pituitary-adrenal axis. Psychoneuroendocrinology. 2001;26(8):761–88.
26. Skelly DT, Hennessy E, Dansereau MA, Cunningham C. A systematic analysis of the peripheral and CNS effects of systemic LPS, IL-1beta, [corrected] TNF-alpha and IL-6 challenges in C57BL/6 mice. PLoS One. 2013;8(7), e69123.
27. Kent S, Bluthe RM, Dantzer R, Hardwick AJ, Kelley KW, Rothwell NJ, et al. Different receptor mechanisms mediate the pyrogenic and behavioral effects of interleukin 1. Proc Natl Acad Sci U S A. 1992;89(19):9117–20.
28. Cunningham C, Wilcockson DC, Campion S, Lunnon K, Perry VH. Central and systemic endotoxin challenges exacerbate the local inflammatory response and increase neuronal death during chronic neurodegeneration. J Neurosci. 2005;25(40):9275–84.
29. Richwine AF, Sparkman NL, Dilger RN, Buchanan JB, Johnson RW. Cognitive deficits in interleukin-10-deficient mice after peripheral injection of lipopolysaccharide. Brain Behav Immun. 2009;23(6):794–802.
30. Swiergiel AH, Dunn AJ. The roles of IL-1, IL-6, and TNFalpha in the feeding responses to endotoxin and influenza virus infection in mice. Brain Behav Immun. 1999;13(3):252–65.
31. Swiergiel AH, Dunn AJ. Distinct roles for cyclooxygenases 1 and 2 in interleukin-1-induced behavioral changes. J Pharmacol Exp Ther. 2002;302(3):1031–6.
32. Teeling JL, Cunningham C, Newman TA, Perry VH. The effect of non-steroidal anti-inflammatory agents on behavioural changes and cytokine production following systemic inflammation: implications for a role of COX-1. Brain Behav Immun. 2010;24(3):409–19.
33. O'Connor JC, Andre C, Wang Y, Lawson MA, Szegedi SS, Lestage J, et al. Interferon-gamma and tumor necrosis factor-alpha mediate the upregulation of indoleamine 2,3-dioxygenase and the induction of depressive-like behavior in mice in response to bacillus Calmette-Guerin. J Neurosci. 2009;29(13):4200–9.

34. O'Connor JC, Lawson MA, Andre C, Moreau M, Lestage J, Castanon N, et al. Lipopolysaccharide-induced depressive-like behavior is mediated by indoleamine 2,3-dioxygenase activation in mice. Mol Psychiatry. 2009;14(5):511–22.

35. Kleinnijenhuis J, van Crevel R, Netea MG. Trained immunity: consequences for the heterologous effects of BCG vaccination. Trans R Soc Trop Med Hyg. 2015;109(1):29–35.

36. Moreau M, Andre C, O'Connor JC, Dumich SA, Woods JA, Kelley KW, et al. Inoculation of Bacillus Calmette-Guerin to mice induces an acute episode of sickness behavior followed by chronic depressive-like behavior. Brain Behav Immun. 2008;22(7):1087–95.

37. O'Connor JC, Lawson MA, Andre C, Briley EM, Szegedi SS, Lestage J, et al. Induction of IDO by bacille Calmette-Guerin is responsible for development of murine depressive-like behavior. J Immunol. 2009;182(5):3202–12.

38. Campisi J, Hansen MK, O'Connor KA, Biedenkapp JC, Watkins LR, Maier SF, et al. Circulating cytokines and endotoxin are not necessary for the activation of the sickness or corticosterone response produced by peripheral E. coli challenge. J Appl Physiol (1985). 2003;95(5):1873–82.

39. Barrientos RM, Watkins LR, Rudy JW, Maier SF. Characterization of the sickness response in young and aging rats following E. coli infection. Brain Behav Immun. 2009;23(4):450–4.

40. Jurgens HA, Amancherla K, Johnson RW. Influenza infection induces neuroinflammation, alters hippocampal neuron morphology, and impairs cognition in adult mice. J Neurosci. 2012;32(12):3958–68.

41. Puntener U, Booth SG, Perry VH, Teeling JL. Long-term impact of systemic bacterial infection on the cerebral vasculature and microglia. J Neuroinflammation. 2012;9:146.

42. Godbout JP, Chen J, Abraham J, Richwine AF, Berg BM, Kelley KW, et al. Exaggerated neuroinflammation and sickness behavior in aged mice following activation of the peripheral innate immune system. FASEB J. 2005;19(10):1329–31.

43. Deng XH, Bertini G, Xu YZ, Yan Z, Bentivoglio M. Cytokine-induced activation of glial cells in the mouse brain is enhanced at an advanced age. Neuroscience. 2006;141(2):645–61.

44. McLinden KA, Kranjac D, Deodati LE, Kahn M, Chumley MJ, Boehm GW. Age exacerbates sickness behavior following exposure to a viral mimetic. Physiol Behav. 2012;105(5):1219–25.

45. Manos PJ, Wu R. The duration of delirium in medical and postoperative patients referred for psychiatric consultation. Ann Clin Psychiatry. 1997;9(4):219–26.

46. Cunningham C, Maclullich AM. At the extreme end of the psychoneuroimmunological spectrum: delirium as a maladaptive sickness behaviour response. Brain Behav Immun. 2013;28:1–13.

47. Davis DH, Skelly DT, Murray C, Hennessy E, Bowen J, Norton S, et al. Worsening cognitive impairment and neurodegenerative pathology progressively increase risk for delirium. Am J Geriatr Psychiatry. 2015;23(4):403–15.

48. Hu X, Chakravarty SD, Ivashkiv LB. Regulation of interferon and Toll-like receptor signaling during macrophage activation by opposing feedforward and feedback inhibition mechanisms. Immunol Rev. 2008;226:41–56.

49. Corona AW, Huang Y, O'Connor JC, Dantzer R, Kelley KW, Popovich PG, et al. Fractalkine receptor (CX3CR1) deficiency sensitizes mice to the behavioral changes induced by lipopolysaccharide. J Neuroinflammation. 2010;7:93.

50. Henry CJ, Huang Y, Wynne AM, Godbout JP. Peripheral lipopolysaccharide (LPS) challenge promotes microglial hyperactivity in aged mice that is associated with exaggerated induction of both pro-inflammatory IL-1beta and anti-inflammatory IL-10 cytokines. Brain Behav Immun. 2009;23(3):309–17.

51. Wohleb ES, Fenn AM, Pacenta AM, Powell ND, Sheridan JF, Godbout JP. Peripheral innate immune challenge exaggerated microglia activation, increased the number of inflammatory CNS macrophages, and prolonged social withdrawal in socially defeated mice. Psychoneuroendocrinology. 2012;37(9):1491–505.

52. Terrando N, Eriksson LI, Ryu JK, Yang T, Monaco C, Feldmann M, et al. Resolving postoperative neuroinflammation and cognitive decline. Ann Neurol. 2011;70(6):986–95.

53. L'Episcopo F, Tirolo C, Testa N, Caniglia S, Morale MC, Impagnatiello F, et al. Switching the microglial harmful phenotype promotes lifelong restoration of subtantia nigra dopaminergic

neurons from inflammatory neurodegeneration in aged mice. Rejuvenation Res. 2011; 14(4):411–24.

54. Lee JW, Lee YK, Yuk DY, Choi DY, Ban SB, Oh KW, et al. Neuro-inflammation induced by lipopolysaccharide causes cognitive impairment through enhancement of beta-amyloid generation. J Neuroinflammation. 2008;5:37.

55. Sheng JG, Bora SH, Xu G, Borchelt DR, Price DL, Koliatsos VE. Lipopolysaccharide-induced-neuroinflammation increases intracellular accumulation of amyloid precursor protein and amyloid beta peptide in APPswe transgenic mice. Neurobiol Dis. 2003;14(1):133–45.

56. Kitazawa M, Oddo S, Yamasaki TR, Green KN, LaFerla FM. Lipopolysaccharide-induced inflammation exacerbates tau pathology by a cyclin-dependent kinase 5-mediated pathway in a transgenic model of Alzheimer's disease. J Neurosci. 2005;25(39):8843–53.

57. Xu Z, Dong Y, Wang H, Culley DJ, Marcantonio ER, Crosby G, et al. Peripheral surgical wounding and age-dependent neuroinflammation in mice. PLoS One. 2014;9(5), e96752.

58. Xu Z, Dong Y, Wang H, Culley DJ, Marcantonio ER, Crosby G, et al. Age-dependent postoperative cognitive impairment and Alzheimer-related neuropathology in mice. Sci Rep. 2014;4:3766.

59. Barrientos RM, Thompson VM, Arnold TH, Frank MG, Watkins LR, Maier SF. The role of hepatic and splenic macrophages in E coli-induced memory impairments in aged rats. Brain Behav Immun. 2015;43:60–7.

60. Hanaa-Mansour A, Hassan WA, Georgy GS. Dexamethazone protects against Escherichia coli induced sickness behavior in rats. Brain Res. 1630;2016:198–207.

61. Kelley KW, O'Connor JC, Lawson MA, Dantzer R, Rodriguez-Zas SL, McCusker RH. Aging leads to prolonged duration of inflammation-induced depression-like behavior caused by Bacillus Calmette-Guerin. Brain Behav Immun. 2013;32:63–9.

62. McManus RM, Higgins SC, Mills KH, Lynch MA. Respiratory infection promotes T cell infiltration and amyloid-beta deposition in APP/PS1 mice. Neurobiol Aging. 2014; 35(1):109–21.

63. Ebert S, Goos M, Rollwagen L, Baake D, Zech WD, Esselmann H, et al. Recurrent systemic infections with Streptococcus pneumoniae do not aggravate the course of experimental neurodegenerative diseases. J Neurosci Res. 2010;88(5):1124–36.

64. Holmes C, Cunningham C, Zotova E, Woolford J, Dean C, Kerr S, et al. Systemic inflammation and disease progression in Alzheimer disease. Neurology. 2009;73(10):768–74.

65. Toovey S, Jick SS, Meier CR. Parkinson's disease or Parkinson symptoms following seasonal influenza. Influenza Other Respir Viruses. 2011;5(5):328–33.

66. Barnes LL, Capuano AW, Aiello AE, Turner AD, Yolken RH, Torrey EF, et al. Cytomegalovirus infection and risk of Alzheimer disease in older black and white individuals. J Infect Dis. 2015;211(2):230–7.

67. Itzhaki RF, Klapper P. Cytomegalovirus: an improbable cause of Alzheimer disease. J Infect Dis. 2014;209(6):972–3.

68. Koch S, Solana R, Dela Rosa O, Pawelec G. Human cytomegalovirus infection and T cell immunosenescence: a mini review. Mech Ageing Dev. 2006;127(6):538–43.

69. Weng NP. Aging of the immune system: how much can the adaptive immune system adapt? Immunity. 2006;24(5):495–9.

70. Weiskopf D, Weinberger B, Grubeck-Loebenstein B. The aging of the immune system. Transpl Int. 2009;22(11):1041–50.

71. Ziv Y, Ron N, Butovsky O, Landa G, Sudai E, Greenberg N, et al. Immune cells contribute to the maintenance of neurogenesis and spatial learning abilities in adulthood. Nat Neurosci. 2006;9(2):268–75.

72. Brynskikh A, Warren T, Zhu J, Kipnis J. Adaptive immunity affects learning behavior in mice. Brain Behav Immun. 2008;22(6):861–9.

73. Wolf SA, Steiner B, Akpinarli A, Kammertoens T, Nassenstein C, Braun A, et al. CD4-positive T lymphocytes provide a neuroimmunological link in the control of adult hippocampal neurogenesis. J Immunol. 2009;182(7):3979–84.

74. Van der Perren A, Macchi F, Toelen J, Carlon MS, Maris M, de Loor H, et al. FK506 reduces neuroinflammation and dopaminergic neurodegeneration in an alpha-synuclein-based rat model for Parkinson's disease. Neurobiol Aging. 2015;36(3):1559–68.

75. Yoshiyama Y, Higuchi M, Zhang B, Huang SM, Iwata N, Saido TC, et al. Synapse loss and microglial activation precede tangles in a P301S tauopathy mouse model. Neuron. 2007;53(3):337–51.

76. Dineley KT, Hogan D, Zhang WR, Taglialatela G. Acute inhibition of calcineurin restores associative learning and memory in Tg2576 APP transgenic mice. Neurobiol Learn Mem. 2007;88(2):217–24.

77. Butovsky O, Koronyo-Hamaoui M, Kunis G, Ophir E, Landa G, Cohen H, et al. Glatiramer acetate fights against Alzheimer's disease by inducing dendritic-like microglia expressing insulin-like growth factor 1. Proc Natl Acad Sci U S A. 2006;103(31):11784–9.

78. Laurie C, Reynolds A, Coskun O, Bowman E, Gendelman HE, Mosley RL. CD4+ T cells from Copolymer-1 immunized mice protect dopaminergic neurons in the 1-methyl-4-phenyl-1,2,3,6-tetrahydropyridine model of Parkinson's disease. J Neuroimmunol. 2007;183(1-2):60–8.

79. Reynolds AD, Banerjee R, Liu J, Gendelman HE, Mosley RL. Neuroprotective activities of CD4+CD25+ regulatory T cells in an animal model of Parkinson's disease. J Leukoc Biol. 2007;82(5):1083–94.

80. Yong J, Lacan G, Dang H, Hsieh T, Middleton B, Wasserfall C, et al. BCG vaccine-induced neuroprotection in a mouse model of Parkinson's disease. PLoS One. 2011;6(1), e16610.

81. Lacan G, Dang H, Middleton B, Horwitz MA, Tian J, Melega WP, et al. Bacillus Calmette-Guerin vaccine-mediated neuroprotection is associated with regulatory T-cell induction in the 1-methyl-4-phenyl-1,2,3,6-tetrahydropyridine mouse model of Parkinson's disease. J Neurosci Res. 2013;91(10):1292–302.

82. Kosloski LM, Kosmacek EA, Olson KE, Mosley RL, Gendelman HE. GM-CSF induces neuroprotective and anti-inflammatory responses in 1-methyl-4-phenyl-1,2,3,6-tetrahydropyridine intoxicated mice. J Neuroimmunol. 2013;265(1-2):1–10.

83. Reynolds AD, Stone DK, Hutter JA, Benner EJ, Mosley RL, Gendelman HE. Regulatory T cells attenuate Th17 cell-mediated nigrostriatal dopaminergic neurodegeneration in a model of Parkinson's disease. J Immunol. 2010;184(5):2261–71.

84. Sanchez-Guajardo V, Annibali A, Jensen PH, Romero-Ramos M. Alpha-Synuclein vaccination prevents the accumulation of Parkinson disease-like pathologic inclusions in striatum in association with regulatory T cell recruitment in a rat model. J Neuropathol Exp Neurol. 2013;72(7):624–45.

85. Toly-Ndour C, Lui G, Nunes MM, Bruley-Rosset M, Aucouturier P, Dorothee G. MHC-independent genetic factors control the magnitude of CD4+ T cell responses to amyloid-beta peptide in mice through regulatory T cell-mediated inhibition. J Immunol. 2011;187(9):4492–500.

86. Jagger A, Shimojima Y, Goronzy JJ, Weyand CM. Regulatory T cells and the immune aging process: a mini-review. Gerontology. 2014;60(2):130–7.

87. Hutter-Saunders JA, Gendelman HE, Mosley RL. Murine motor and behavior functional evaluations for acute 1-methyl-4-phenyl-1,2,3,6-tetrahydropyridine (MPTP) intoxication. J Neuroimmune Pharmacol. 2012;7(1):279–88.

88. Saunders JA, Estes KA, Kosloski LM, Allen HE, Dempsey KM, Torres-Russotto DR, et al. CD4+ regulatory and effector/memory T cell subsets profile motor dysfunction in Parkinson's disease. J Neuroimmune Pharmacol. 2012;7(4):927–38.

89. Larbi A, Fulop T. From "truly naive" to "exhausted senescent" T cells: when markers predict functionality. Cytometry A. 2014;85(1):25–35.

90. Pellicano M, Larbi A, Goldeck D, Colonna-Romano G, Buffa S, Bulati M, et al. Immune profiling of Alzheimer patients. J Neuroimmunol. 2012;242(1–2):52–9.

91. Butchart J, Brook L, Hopkins V, Teeling J, Puntener U, Culliford D, et al. Etanercept in Alzheimer disease: a randomized, placebo-controlled, double-blind, phase 2 trial. Neurology. 2015;84(21):2161–8.

Frailty and Ageing

9

Thomas A. Jackson, Daisy Wilson, and Carolyn A. Greig

Abstract

Frailty is a common condition affecting older adults, but is not an inevitable part of the ageing process. Frailty is a method of describing the observed heterogeneity in function, disability and health among older people. Importantly what frailty identification allows is the prediction of risk of adverse events such as mortality or disability. Frailty is defined as a syndrome with multiple causes and contributors that is characterised by diminished strength, endurance and reduced

T.A. Jackson (✉)
Institute of Inflammation and Ageing, University of Birmingham,
University of Birmingham Research Laboratories, Queen Elizabeth Hospital,
Mindelsohn Way, Edgbaston, Birmingham B15 2WD, UK

Queen Elizabeth Hospital, University Hospitals Birmingham Foundation NHS Trust,
Mindelsohn Way, Edgbaston, Birmingham B15 2WD, UK
e-mail: t.jackson@bham.ac.uk

D. Wilson
Institute of Inflammation and Ageing, University of Birmingham,
University of Birmingham Research Laboratories, Queen Elizabeth Hospital,
Mindelsohn Way, Edgbaston, Birmingham B15 2WD, UK

MRC-Arthritis Research UK Centre for Musculoskeletal Ageing Research,
University of Birmingham, Edgbaston, Birmingham B15 2TT, UK
e-mail: wilsonDV@adf.bham.ac.uk

C.A. Greig
MRC-Arthritis Research UK Centre for Musculoskeletal Ageing Research,
University of Birmingham, Edgbaston, Birmingham B15 2TT, UK

School of Sport, Exercise and Rehabilitation Sciences, University of Birmingham,
Edgbaston, Birmingham B15 2TT, UK
e-mail: c.a.greig@bham.ac.uk

© Springer International Publishing Switzerland 2017
V. Bueno et al. (eds.), *The Ageing Immune System and Health*,
DOI 10.1007/978-3-319-43365-3_9

physiological function that increases an individual's vulnerability for developing increased dependency and/or death. Frailty is important clinically as older people with frailty present to healthcare services atypically with frailty syndromes such as falls and delirium. Frail older adults may benefit from assessment by specialist geriatricians and a wider multi-disciplinary team.

Frailty is associated with changes to the immune system, importantly the presence of a pro-inflammatory environment and changes to both the innate and adaptive immune system. Some of these changes have been demonstrated to be present before the clinical features of frailty are apparent suggesting the presence of potentially modifiable mechanistic pathways. To date, exercise programme interventions have shown promise in the reversal of frailty and related physical characteristics, but there is no current evidence for successful pharmacological intervention in frailty. Multicomponent interventions such as those based on Comprehensive Geriatric Assessment have also demonstrated some reversal of frailty.

There are numerous knowledge gaps within the literature. A clearer understanding of the fundamental science underpinning frailty is vital to allow for the development of successful interventions

Keywords
Frailty • Sarcopenia • Immunity • Aged • Comprehensive geriatric assessment • IL-6 • IL-10 • Muscle • Immunesenescence • Inflammageing

9.1 Introduction, Definitions and Operationalisation

The Oxford English dictionary defines frailty as 'the condition of being weak and delicate'. In the clinical setting, a number of key terms including 'decreased reserve', 'increased vulnerability' and 'declines across multiple physiological systems' are embedded within a number of published definitions [1–3].

More recently, the Frailty Consensus Conference highlighted the need to specify physical frailty as a medical syndrome and proposed it be defined as "a medical syndrome with multiple causes and contributors that is characterized by diminished strength, endurance, and reduced physiologic function that increases an individual's vulnerability for developing increased dependency and/or death" [4]. Frailty is associated with increased risk of morbidity and mortality [5]. By defining frailty, the observed heterogeneity in function, ability and health status of older adults can be described.

There is no current consensus on a single *operational* definition with respect to assessing frailty status. The most commonly reported methods are based either on a frailty 'phenotype', i.e., problems within two or more domains of physical function [5] or on a frailty 'index' based upon a proportion of accumulated deficits [3]. Both have been widely tested for validity but not for reliability and neither instrument is regarded as a 'gold standard' [6]. The Fried frailty phenotype is based on five components: shrinking (unintentional weight loss), weakness (low grip strength), poor endurance and energy (self-report exhaustion), slowness (slow walking speed) and low physical activity. Individuals meeting ≥3 of these five operational criteria are characterized as frail; those who meet one or two criteria are pre-frail, i.e., they are at risk of progressing to a frail state while those who do not meet any of the operational criteria are classified as non-frail [5]. The Frailty Index is based on the accumulation of up to 70 variables or 'deficits'. Health deficits can be symptoms, diagnosed diseases, laboratory abnormalities, or disability. Deficits must become more prevalent with age, not saturate too early and cover a range of health systems. Subsequently the more deficits a person accumulates, the more likely that person is to be frail [3].

It is suggested that these assessment instruments should not be considered as alternatives, but rather as complementary as they serve different purposes. Thus the frailty phenotype may have utility in identifying older adults at risk of disability whereas the Frailty Index acts as an objective marker of deficit accumulation and/ or assessment of the efficacy of an intervention [7]. Equally, in terms of measuring outcomes, the frailty phenotype is a categorical classification (not frail, pre-frail, or frail) whereas the frailty index gives a continuous measured classification between zero and one (Fig. 9.1, Table 9.1).

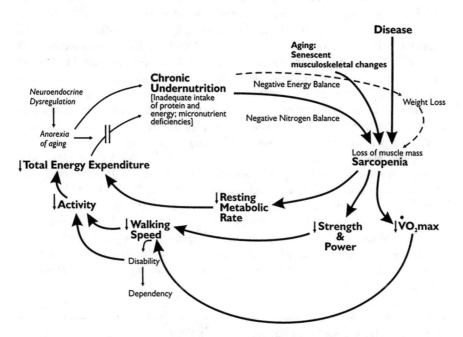

Fig. 9.1 Cycle of frailty as hypothesised in the original frailty phenotype paper. This demonstrates the inter-relationship between disease and symptoms of frailty, which in turn worsen disease, thus creating a cycle of accumulating deficits. Age related changes to the immune system would impact on the degree of affect disease has on this cycle. Reproduced with permission from: Linda P. Fried et al. J Gerontol A Biol Sci Med Sci 2001;56:M146–M157

Table 9.1 The main differences between the frailty phenotype model and the deficit accumulation model

Frailty phenotype model	Deficit accumulation model (frailty index)
Signs, symptoms	Diseases, activities of daily living, results of a clinical evaluation
Possible before a clinical assessment	Only possible after a comprehensive clinical assessment
Categorical variable	Continuous variable
Pre-defined set of criteria	Unspecified set of criteria
Possible to define a 'pre-frailty' syndrome	Frailty as an accumulation of deficits
Meaningful results potentially restricted to non-disabled older persons	Meaningful results in every individual, independently of functional status or age

Adapted from: Cesari M et al. The frailty phenotype and the frailty index: different instruments for different purposes. Age Ageing. 2014;43(1):10–12

9.1.1 Prevalence

Although physical frailty is not exclusively a phenomenon of old age, its prevalence increases with advancing age and is most prevalent in persons over 85 years [8]. A recent systematic review which included 21 community-based studies of 61,500 participants aged 65 years or older reported a global prevalence of frailty assessed using a definition (mostly the Fried Frailty phenotype) of 4.0–59.1 % (weighted mean 9.9 % in those studies utilising a physical frailty definition and mean 7.9 % using the Fried Frailty phenotype). The weighted mean was higher (13.6 %) in studies which employed a broader phenotypic definition. The prevalence of frailty in those over 85 years of age was 26.1 % (the assumption is that this value would be higher in older adults in institutionalised care). Women were almost twice as likely as men to be frail (9.6 % versus 5.2 %; [8]). The reported gender difference confirms a previous report (8 % versus 5 %; [5]) and may be due to a longer life expectancy in women and/ or more years of terminal dependency compared with men [8]. A more recent European-wide study, the Survey of Health, Aging and Retirement in Europe (SHARE) which included more than 85,000 individuals aged over 65 years reported an average prevalence of 17 % (range 5.8–27.3 %), [9]. According to another review, the mean prevalence of frailty in Latin America and Caribbean older people is even higher, 22–35 % in men and 30–48 % in women [1, 10]. In Japan, the prevalence of frailty in a community-dwelling older people was 24.3 % for men and 32.4 % for women, and increased for age ≥80 years to 45.3 % for men and 50 % for women [11].

9.2 Clinical Utility of Identifying Frailty

The clinical utility of identifying frailty is related to the fact that frailty is strongly associated with increased health care service use [12]. Initially a concept in the geriatric literature, frailty has now been explored in both the community and hospital settings. Frailty has been recognised as a risk factor for adverse events in both community populations [13] and older people admitted to hospital [14]. Specialist hospital services such as surgery [15], hip fractures [16] and oncology [17] have now begun to recognise frailty as an important predictor of mortality and morbidity. In high risk healthcare settings such as care homes, there are very high levels of frailty [18]. As the population ages and healthcare systems begin to adjust to a different demand, then frailty can be used as an important construct to guide changes in healthcare resources [19].

The clinical significance of frailty lies in the consequences of acute illness on a frail person. Older people with frailty are common users of emergency departments because frailty may predispose them to acute illness due to reduced physiological reserve [20]. Acute illness in a frail adult often presents atypically (differently than in younger adults) and this has given rise to the frailty syndromes of delirium, falls, immobility and susceptibility to drug side effects [21]. Heightened awareness and identification of frailty in clinical practice can then allow implementation of geriatric medicine focussed management as well as specific interventions which will be discussed later. In practice, acute illness in a frail person results in a disproportionate change in a frail person's functional ability when faced with a relatively minor physiological stressor, associated with a prolonged recovery time or early tipping point [22]. This is illustrated in Fig. 9.2 [23].

The identification of frailty as part of routine medical practice has been advocated with the development of a number of screening tools to allow clinicians to

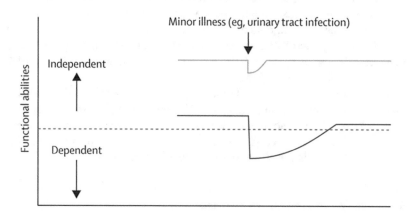

Fig. 9.2 Vulnerability of frail older people to a sudden change in functional ability after a minor illness. The *upper line* represents a fit older individual who, after a minor stressor event such as an infection, has a small deterioration in function and then returns to full function. The *lower line* represents a frail older individual who, after a similar stressor event, undergoes a larger deterioration, which may manifest as functional dependency, and who fails to return to previous function over a longer recovery period. The *horizontal dashed line* represents the threshold between independence and dependence. Reprinted from The Lancet, Vol.381, Clegg, A; Young, J; Iliffe, S; Rikkert, Mo; Rockwood, K. Frailty in Elderly People, 752–767, Copyright (2013), with permission from Elsevier

rapidly assess frailty [24]. The best described of these is the PRISMA-7 questionnaire, a short self-reported questionnaire, which is recommended as the most appropriate screening tool by the British Geriatrics Society [25]. This allows rapid identification of frailty and has reasonable diagnostic test accuracy with a sensitivity of 83 % and a specificity of 83 % when compared to a comprehensive geriatric assessment identifying frailty by the Fried phenotype [26]. A simple functional test, such as a timed walk or timed get up and go test has also been suggested as a good screening tool for frailty [26].

Clinical operationalisation of the deficit accumulation model of frailty has been seen with the Rockwood Clinical Frailty Scale [3], as well as a frailty indices based on routine laboratory tests [27], biomarkers of ageing [28] and routinely collected primary healthcare data [29]. All these scales accurately predict the risk of adverse events such as mortality or new institutionalisation.

Comprehensive Geriatric Assessment (CGA) is a multidisciplinary, multidimensional assessment, led by geriatricians [30]. CGA is designed as both a diagnostic tool and a tool to develop treatment plans. The key objective of CGA is to diagnose specific geriatric conditions. CGA involves the detailed assessment of the patient by a geriatrician, physiotherapist, occupational therapist, social worker and others professionals. This includes physical examination, medical history, and functional assessments of the patient as well as assessment of their home environment and psychosocial support. The information from these assessments form the management plan which involves specific treatment goals that are patient and carer focused. The primary aims of CGA are to improve both physical and psychological health, reduce or shorten hospital stay, reduce institutionalisation and improve quality of life [31]. A meta-analysis of CGA in hospitals in 2011 identified 22 randomised trials with a total of 10,315 older patients included. This reported that older patients receiving CGA in hospital increased the chance of being alive or in their own home at six months by 25 % (OR 1.25 95 % CI 1.11–1.42; P<0.001) [32].

CGA also improves outcomes in other settings such as stroke medicine [33], hip fracture management [34], surgical care [35], care homes [36] and oncology [37]. However, to date there is no direct evidence that CGA prevents or improves frailty, only the consequences of frailty. Research investigating the improvement of frailty with CGA is challenging due to various methodological issues [38].

9.3   Frailty and Immunesenescence

The role played by the immune system in frailty is believed to be fundamental in the development and maintenance of a frailty state, but it remains mostly uncharacterised. The role of senescence and inflammageing has been discussed in detail earlier in Chap. 1 and in this chapter we shall focus on findings directly related to frailty.

9.3.1   Models

There has been little fundamental mechanistic research investigating the immune system in frailty in its truest sense. Therefore, it is necessary to consider models of frailty and draw conclusions where possible. Some of the models of frailty that have been utilised in research are discussed below.

Population based studies have demonstrated that the incidence of infection and subsequent mortality is higher in populations of frail people. Nursing homes represent a population where frailty can be assumed without direct measurement. The prevalence of pneumonia in a nursing home population is 30 times higher than the general population [39, 40]. Population based studies have identified that frailty is associated with inflammageing and immunesenescence [41–44]. As HIV could lead to the same characteristics of immunesenescence in older individuals, the identification of a subpopulation of HIV patients with frailty has led to comparison cohort studies between HIV patients with frailty and those without [45]. These studies focus upon the adaptive immune system and T cells in particular. Conversely, centenarians have been employed as a model of successful ageing [46, 47].

Sarcopenia has been suggested as a human model of physical frailty. Sarcopenia as defined by EWGSOP is a syndrome characterised by progressive and generalised loss of skeletal muscle mass and strength with the risk of adverse outcomes such as physical disability, poor quality of life and death [48]. It is a more tangible concept than frailty and is easier to utilise in a research setting. Severe sarcopenia as defined by EWGSOP is considered pre-frailty as defined by the frailty phenotype. This suggests it could be either the physical component or a precursor to frailty. The prevalence of sarcopenia in frailty is unknown and this represents an important knowledge gap. However, the most common frailty phenotype parameters are slow gait speed and weakness which are also sarcopenia defining criteria [49].

Murine models have been developed to investigate sarcopenia and frailty using either cell signalling pathway knock out models or accelerated ageing models. Cell signalling pathway knock out models display changes in muscle fibre size, growth and the proportions of type 1 and 2 muscle fibres. In addition to these changes accelerated ageing models also display additional features of ageing such as osteoporosis and alopecia [50]. One murine model, the IL-10 knock out mouse, has been reported as a specific murine model of frailty [51] and a valid murine frailty index has been developed to improve murine frailty research [52]. These models offer an excellent and unique opportunity to study frailty in vivo but lack the sophistication and subtleties of frailty in humans.

9.3.2   Inflammageing

The concept of inflammageing, a state of discrete chronic inflammation, has been discussed in depth in Chap. 1. In brief, it is characterised by an age related immune imbalance with increased pro-inflammatory cytokines such as IL-6, CRP, TNFα and decreased IL-10, an anti-inflammatory cytokine.

A positive association has been demonstrated between IL-6 and CRP with frailty on multiple occasions in different frail populations [41–43, 53]. High levels of IL-6 predict disability and mortality, which are significant outcomes in frailty, in older community dwelling adults [54–56]. PBMCs from frail older adults produce greater amounts of IL-6 on stimulation compared to non-frail counterparts [57].

Inflammatory and immunological biomarkers of frailty have been used in a variety of settings to predict poor outcomes [28, 58] The Biomarker Frailty Index incorporates markers of inflammation and immunesenescence as well as haematological and genetic markers. This model is associated with mortality and at low levels of frailty is better than the frailty index at predicting mortality [28].

Investigation of the individual components of sarcopenia and frailty has demonstrated an association with inflammageing and inflammageing predating a loss of physical function. Cross-sectional studies have demonstrated both an association of high levels of IL-6 and TNFα with low muscle mass and strength [59] and low levels of CRP with high grip strength [60]. Longitudinal studies have shown that higher levels of IL-6 predict reduction in fat free mass, a hallmark of sarcopenia [61]. Higher levels of IL-6 and CRP at baseline equate to a two- to threefold greater risk of losing more than 40 % of grip strength over a 3-year follow up [62]. TNFα mRNA and protein levels are higher in frail older adults. It was also demonstrated that this could be reversed with an exercise program, which correlated with an increase in muscle strength [63].

The frailty murine model, IL-10 knock out mouse, has increased levels of IL-6 at 50 weeks and phenotypically displays increased muscle weakness and decreased activity levels in comparison to the wild type [51]. In murine models the exogenous administration of TNFα results in anorexia and weight loss in the mouse [64, 65].

In centenarians IL-6 −174 GG genotype, associated with higher plasma levels of IL-6, is under represented and IL-10 −1082 CC, associated with higher plasma levels of IL-10, are over-represented [46].

The limited data available demonstrates that frailty is associated with a state of chronic inflammation. There is also evidence that inflammageing predates a diagnosis of frailty suggesting a causative role.

9.3.3   Innate Immunity and Frailty

A small number of studies have demonstrated a dysregulation of the innate immune system in frailty. Frail adults have raised white cell and neutrophil count. This remains significantly elevated even after adjustment for common confounders [66–68]. Raised neutrophils are also associated with low levels of physical activity and frailty [69]. High white cell count can predict frailty at a ten year follow up [70]. An association has been demonstrated between frailty and the upregulation of monocytic expression of CXCL 10 gene which codes for a potent pro-inflammatory chemokine [71].

9.3.4 Adaptive Immunity and Frailty

A small number of studies have described the dysregulation of the adaptive immune system in frailty, but the results have been conflicting

A prospective review of response to vaccination showed a reduction in adaptive immune function in frail individuals [72]. However, the study did not attempt to investigate the individual roles of T and B cells in poor response to vaccination. Initial work into frailty in HIV patients demonstrated higher levels of frailty in HIV positive patients compared to non-HIV infected age matched controls [73] and subsequently that CD4+ T cell count can predict the development of frailty in HIV positive men independent of HAART and plasma viral load [45]. Research into T cell subset expression in frailty has revealed conflicting results, particularly in the investigation of CD4:CD8 ratio which has both been positively [68, 74] and inversely associated with frailty [75, 76]. The association of CMV with frailty is also unclear. Frailty has been associated with higher titres of CMV IgG [77, 78] and also higher levels of CD8+CD28− T cells [74, 76]. CD8+CD28− T cells are highly associated with seropositivity to persistent infections such as CMV. However, the extremely high prevalence of CMV in all older adults make these data difficult to interpret. Only one study has investigated the B cells of a frail population and showed that with increasing age the diversity of the B cell population decreases and this correlates with poor health [79].

To date there is limited research describing a direct relationship between frailty, and immunesenescence or inflammageing. However, initial results have demonstrated an association of frailty with a state of chronic inflammation and dysregulation of both the innate and adaptive immune system. It has also established that immunesenescence and inflammageing predate a diagnosis of frailty suggesting a causative role.

9.4 Interventions for Frailty

Given the general impact of frailty on the individual and society, it is clear that interventions, both preventive and therapeutic, are required. Interventional studies where a change in frailty status is the primary outcome are limited. Due to the lack of research with these clear outcomes, it is necessary to consider interventional studies using measurements of an element of frailty such as muscle strength, or walking speed as a primary outcome. The interventions are often specifically exercise and nutrition based, or pharmacological. However some trials use multifactorial interventions and the use of the Comprehensive Geriatric Assessment.

9.4.1 Exercise and Nutritional Based Interventions

When considering exercise and nutritional based interventions in frailty there is conflicting evidence within the available literature. To some extent this may be due to the variability in outcomes and inclusion criteria. Clinical studies have both analysed the independent and combined effect of exercise programmes and nutrition on frailty and the components of frailty.

A recent meta-analysis and four individual systematic reviews have found beneficial evidence of exercise programmes on selected physical and functional ability [80–83]. Improvements have been demonstrated in gait speed [80, 84], balance [80], performance in activities of daily living (ADL) [80, 85] and the Short Physical Performance Battery (SPPB) [84]. However, there was substantial variability in the method of frailty assessment in these studies, with the meta-analysis including only participants classified as frail according to a recognised frailty definition (eight RCTs, 1068 participants classified as frail according to Fried Frailty Index, Speechley and Tinetti criteria and Edmonton Frail Scale) [80]. A recent systematic review reported that exercise programmes appeared to be more effective in studies which did not assess frailty using an operational definition. The authors suggested that the participants in these studies may have been more responsive to exercise due to better baseline health or actually being non-frail [81]. This suggests that exercise may be less effective once clinical frailty is evident. A further review found that although five out of six exercise intervention studies reported benefit in at least one performance measure in 'frail' individuals [83], the one study reporting no benefit was the only one assessing frailty using an operational definition [85]. This adds evidence that exercise interventions may have no positive effect in operationally defined frail individuals. Another recent review of 19 RCTs that included recruited participants defined as, or considered as, frail reported variable benefits of an exercise programme; evidence of benefit was shown for gait speed and short physical performance battery but no consistent effect on ADL or balance was demonstrated [84].

Epidemiological studies have shown associations between protein intake and frailty: The InCHIANTI study reported that the risk of frailty (Fried frailty criteria) was twice as high in those in the lowest quintile of protein intake compared with those with higher intakes [86]. These data were supported by a later study of 24,417 participants (frailty-free at baseline) in the Women's Health Initiative Observational Study which showed that a higher baseline protein intake at baseline was associated with a reduction of frailty risk after 3 years of follow up [87]. A more recent cross-sectional study of 194 community dwelling adults older than 75 years reported an association between frailty (Fried frailty criteria) and distribution of daily protein intake [88]. Frailer individuals did not eat less protein compared with non-frail counterparts (median daily intake 1.07 g kg^{-1}) but they had a more uneven distribution of protein. There are very few studies solely using nutrition as an intervention: one study of 87 older adults classified as frail according to the Fried frailty criteria and taking a commercial supplement containing additional calories, protein, and essential amino acids twice daily for 12 weeks, showed improvements in selected functional outcomes compared with controls: improved physical function and timed get up and go, a stable SPPB score compared with a decrease SPPB score in the control group and attenuated reduction in gait speed when compared with the control group [89]. However, no benefit in muscle strength or function from a protein-energy supplement was reported in an earlier trial that did not use a standardised assessment of frailty [90]. In a further trial participants identified as frail on the basis of physical inactivity and weight loss showed no significant effects on fitness or disability measures [91]. The same study included an exercise-only group and showed benefit in functional performance. One of the major methodological challenges in nutritional supplementation studies with frail adults is attempting to control for baseline nutritional status, weight and body composition, as well as accounting for adherence to supplementations in the real world. This may explain the lack of major effects to date.

Other studies have combined nutrition with physical activity and exercise training [92]. A recent RCT in 62 institutionalised older adults classified as frail using the Fried frailty criteria reported that protein supplementation combined with resistance exercise training increased muscle mass compared with a training-only group [93]. In the Singapore Frailty Intervention Trial, 246 frail older adults were randomised into five groups: 48 received just physical training and 49 received just nutritional supplementation for 6 months. The other three groups received cognitive training, a combined intervention and usual care respectively. At both 6 and 12 month follow-up a significant reduction in frailty scores compared with control was shown for both interventions when delivered individually as well as in combination with cognitive training [92]. A pilot RCT study of 117 pre-frail and frail Taiwanese older adults reports that those randomised to a 3-month combination of nutritional consultation and exercise training showed a short-term beneficial effect on frailty status at 3 months, but not at 6 or 12 months compared with those receiving the interventions individually [94]. The nutritional component in this study however was educational in nature rather than a direct intervention using dietary modification or supplementation. Results from the LIFE-P trial comparing a physical activity intervention with a successful lifestyle (educational) intervention in 424 community dwelling older adults showed a significant difference in frailty (Fried criteria) prevalence after 12 months in the physical activity group [95].

Read together these reports highlight a major unmet need for trials that examine exercise interventions, and nutritional supplements to specifically improve physical function in frail older adults. With the recent recommendation to increase protein intake in older age [96], the relative lack of nutrition and exercise intervention study data to improve frailty status is surprising.

9.4.2 Pharmacological Interventions

To date there is no clear evidence that pharmacological interventions improve or ameliorate frailty.

ACE inhibitors, targeting the renin-angiotensin-aldostrerone axis have been shown to have no effect on incident frailty in a cohort of older women (aged 65–79 years at baseline) followed up for 3 years [97]. However, a single randomised control trial of perindopril used over 20 weeks showed an improvement in 6 min walking distances in 130 older adults (mean age 79 years) with pre-existing mobility problems [98].

Vitamin D is a potential candidate for pharmacological management of frailty given its effects on skeletal muscle, osteoporosis, and falls prevention. However, studies to date have demonstrated conflicting results. Higher serum 25-hydroxyvitamin D levels are associated with a lower prevalence of frailty in older men and women in observational studies [99, 100]. Vitamin D supplementation has been shown to improve functional performance in older fallers with low serum 25-hydroxyvitamin D [101], but a recent real world RCT of high dose vitamin D supplementation in older fallers with both reduced and normal serum 25-hydroxyvitamin D demonstrated no improvement in lower limb function and a surprising increase in falls at 6 and 12 months [102].

Testosterone supplementation as a pharmacological intervention has shown conflicting results. Older men with low plasma testosterone treated for 36 months with transdermal testosterone, showed a significant improvement in lean muscle mass, but this did not translate to improvements in function [103]. Similar results were seen in older men with low testosterone and frailty given transdermal testosterone for 12–24 months [104]. However, improvements in hand grip strength and timed walking were found in older men treated with intramuscular testosterone therapy [105]. A recent large RCT of 790 older men with low serum testosterone given transdermal testosterone for 12 months showed no benefit to walking distance or self-reported vitality [106]. The main concern with testosterone treatment is its potential for cardiovascular side effects.

Replacement of dehydroepiandrosterone (DHEA), a steroid precursor of testosterone is a plausible treatment as levels of DHEA fall with advancing age and lower levels are associated with mortality and frailty [107]. A meta-analysis from 25 trials (n = 1353) of DHEA treatment in older men concluded that DHEA had a small effect on body composition but was not able to demonstrate an effect on function at a mean follow-up of 36 months.

Reduced IGF-1 is associated with frailty components [108], but to date treatment with growth hormone in the somatopause (the age related reduction in circulating growth hormone) leads to adverse events and only a minimal change to body composition and no improvement in function [109].

Of interest, the TNF-α antagonist drugs and other biologicals that have been used to treat diseases such as rheumatoid arthritis show a positive effect on the systemic symptoms of these diseases such as weakness and fatigue [110, 111]. These could be considered components of frailty and as such merit further investigation in a frail population.

Clearly further high quality trials are needed to discover new treatments and a search of the National Library of Medicine clinical trials registration website in January 2016 identified 15 ongoing interventional trials with frailty as an outcome including important trials of multifactorial interventions [112].

9.4.3 Multifactorial Investigations Based on Comprehensive Geriatric Assessment (CGA)

A randomised trial of a multifactorial intervention based explicitly on CGA (medication adjustment, exercise, nutritional support, rehabilitation, social worker consultation and speciality referral) in 310 frail older men and measured using the Fried frailty criteria showed those who received CGA were less likely to deteriorate, but this was not significant (OR = 1.19, 95 % confidence interval = 0.48–3.04, p=0.71).

The Frailty Intervention Trial (FIT) was established to provide a multidisciplinary, multifaceted intervention that specifically targeted individually identified components of the Fried frailty criteria (weight loss, weakness, exhaustion, slowness, and low activity) [113]. It comprised 241 older people identified as frail (positive for three or more of the Fried criteria) and randomised to the intervention or usual healthcare. The intervention was delivered by a multidisciplinary team led by a geriatrician and was akin to CGA. At 12 months there was a lower prevalence of frailty in the intervention group and a reduction in the number of Fried criteria by an average of 0.80 (SD = 1.19) in the intervention group compared with 0.41 (SD = 1.02) in the control group [114]. Adherence to the interventions was the significant limitation (median adherence to each interventions was only 50%), reflecting the difficulties inherent to the implementation of multifactorial interventions in the real world.

9.5 Conclusions

To conclude we have described frailty, a common condition important to both clinicians and fundamental scientists. There remains considerable uncertainty about the concept, definition, and operationalisation of frailty; in particular we have described the differences between the two core theories of frailty, the phenotype model, and the deficit accumulation model. The identification of frailty clinically can signpost older people towards healthcare interventions such as Comprehensive Geriatric Assessment. However, there is still a large research gap in terms of both targeted and multifactorial interventions to prevent, reverse or ameliorate frailty. The study of fundamental mechanisms in the development of frailty shows real promise. Early research has demonstrated immune dysfunction which predates the development of the clinical features of frailty. Thus, further description of these mechanisms and their relationship to frailty and sarcopenia, along with identification of potential pathways amenable to intervention should be a research priority.

References

1. Xue QL. The frailty syndrome: definition and natural history. Clin Geriatr Med. 2011;27(1): 1–15.
2. Campbell AJ, Buchner DM. Unstable disability and the fluctuations of frailty. Age Ageing. 1997;26(4):315–8.
3. Rockwood K, Song X, MacKnight C, Bergman H, Hogan DB, McDowell I, et al. A global clinical measure of fitness and frailty in elderly people. CMAJ. 2005;173(5):489–95.
4. Morley JE, Vellas B, van Kan GA, Anker SD, Bauer JM, Bernabei R, et al. Frailty consensus: a call to action. J Am Med Dir Assoc. 2013;14(6):392–7.
5. Fried LP, Tangen CM, Walston J, Newman AB, Hirsch C, Gottdiener J, et al. Frailty in older adults: evidence for a phenotype. J Gerontol A Biol Sci Med Sci. 2001;56(3):M146–56.
6. Bouillon K, Kivimaki M, Hamer M, Sabia S, Fransson EI, Singh-Manoux A, et al. Measures of frailty in population-based studies: an overview. BMC Geriatr. 2013;13:64.
7. Cesari M, Gambassi G, van Kan GA, Vellas B. The frailty phenotype and the frailty index: different instruments for different purposes. Age Ageing. 2014;43(1):10–2.

8. Collard RM, Boter H, Schoevers RA, Oude Voshaar RC. Prevalence of frailty in community-dwelling older persons: a systematic review. J Am Geriatr Soc. 2012;60(8):1487–92.

9. Michel JP, Cruz-Jentoft AJ, Cederholm T. Frailty, exercise and nutrition. Clin Geriatr Med. 2015;31(3):375–87.

10. Alvarado BE, Zunzunegui MV, Beland F, Bamvita JM. Life course social and health conditions linked to frailty in Latin American older men and women. J Gerontol A Biol Sci Med Sci. 2008;63(12):1399–406.

11. Shinkai S, Yoshida H, Taniguchi Y, Murayama H, Nishi M, Amano H, et al. Public health approach to preventing frailty in the community and its effect on healthy aging in Japan. Geriatr Gerontol Int. 2016;16 Suppl 1:87–97.

12. Rochat S, Cumming RG, Blyth F, Creasey H, Handelsman D, Le Couteur DG, et al. Frailty and use of health and community services by community-dwelling older men: the Concord Health and Ageing in Men Project. Age Ageing. 2010;39(2):228–33.

13. Mitnitski AB, Graham JE, Mogilner AJ, Rockwood K. Frailty, fitness and late-life mortality in relation to chronological and biological age. BMC Geriatr. 2002;2:1.

14. Joosten E, Demuynck M, Detroyer E, Milisen K. Prevalence of frailty and its ability to predict in hospital delirium, falls, and 6-month mortality in hospitalized older patients. BMC Geriatr. 2014;14:1.

15. Makary MA, Segev DL, Pronovost PJ, Syin D, Bandeen-Roche K, Patel P, et al. Frailty as a predictor of surgical outcomes in older patients. J Am Coll Surg. 2010;210(6):901–8.

16. Patel KV, Brennan KL, Brennan ML, Jupiter DC, Shar A, Davis ML. Association of a modified frailty index with mortality after femoral neck fracture in patients aged 60 years and older. Clin Orthop Relat Res. 2014;472(3):1010–7.

17. Hamaker ME, Jonker JM, de Rooij SE, Vos AG, Smorenburg CH, van Munster BC. Frailty screening methods for predicting outcome of a comprehensive geriatric assessment in elderly patients with cancer: a systematic review. Lancet Oncol. 2012;13(10):e437–44.

18. Gordon AL, Franklin M, Bradshaw L, Logan P, Elliott R, Gladman JR. Health status of UK care home residents: a cohort study. Age Ageing. 2014;43(1):97–103.

19. Oliver D. Transforming care for older people in hospital: physicians must embrace the challenge. Clin Med. 2012;12(3):230–4.

20. Salvi F, Morichi V, Grilli A, Lancioni L, Spazzafumo L, Polonara S, et al. Screening for frailty in elderly emergency department patients by using the identification of seniors at risk (ISAR). J Nutr Health Aging. 2012;16(4):313–8.

21. Soong J, Poots AJ, Scott S, Donald K, Bell D. Developing and validating a risk prediction model for acute care based on frailty syndromes. BMJ Open. 2015;5(10):008457.

22. Olde Rikkert MG, Dakos V, Buchman TG, Boer R, Glass L, Cramer AO, et al. Slowing down of recovery as generic risk marker for acute severity transitions in chronic diseases. Crit Care Med. 2016;44(3):601–6.

23. Clegg A, Young J, Iliffe S, Rikkert MO, Rockwood K. Frailty in elderly people. Lancet. 2013;381(9868):752–62.

24. Hoogendijk EO, van der Horst HE, Deeg DJ, Frijters DH, Prins BA, Jansen AP, et al. The identification of frail older adults in primary care: comparing the accuracy of five simple instruments. Age Ageing. 2013;42(2):262–5.

25. Turner G, Clegg A. Best practice guidelines for the management of frailty: a British Geriatrics Society, Age UK and Royal College of General Practitioners report. Age Ageing. 2014;43(6):744–7.

26. Clegg A, Rogers L, Young J. Diagnostic test accuracy of simple instruments for identifying frailty in community-dwelling older people: a systematic review. Age Ageing. 2014;44(1):148–52.

27. Howlett SE, Rockwood MR, Mitnitski A, Rockwood K. Standard laboratory tests to identify older adults at increased risk of death. BMC Med. 2014;12:171.

28. Mitnitski A, Collerton J, Martin-Ruiz C, Jagger C, von Zglinicki T, Rockwood K, et al. Age-related frailty and its association with biological markers of ageing. BMC Med. 2015;13:161.

29. Clegg A, Bates C, Young J, Ryan R, Nichols L, Ann Teale E, et al. Development and validation of an electronic frailty index using routine primary care electronic health record data. Age Ageing. 2016;45(3):353–60.

30. Stuck AE, Iliffe S. Comprehensive geriatric assessment for older adults. BMJ. 2011;343.
31. Welsh TJ, Gordon AL, Gladman JR. Comprehensive geriatric assessment – a guide for the non-specialist. Int J Clin Pract. 2014;68(3):290–3.
32. Ellis G, Whitehead MA, Robinson D, O'Neill D, Langhorne P. Comprehensive geriatric assessment for older adults admitted to hospital: meta-analysis of randomised controlled trials. BMJ. 2011;343:d6553.
33. Stroke Unit Trialists C. Organised inpatient (stroke unit) care for stroke. Cochrane Database Syst Rev. 2013;9:CD000197.
34. Prestmo A, Hagen G, Sletvold O, Helbostad JL, Thingstad P, Taraldsen K, et al. Comprehensive geriatric care for patients with hip fractures: a prospective, randomised, controlled trial. Lancet. 2015;385(9978):1623–33.
35. Partridge JS, Harari D, Martin FC, Dhesi JK. The impact of pre-operative comprehensive geriatric assessment on postoperative outcomes in older patients undergoing scheduled surgery: a systematic review. Anaesthesia. 2014;69 Suppl 1:8–16.
36. Stuck AE, Egger M, Hammer A, Minder CE, Beck JC. Home visits to prevent nursing home admission and functional decline in elderly people: systematic review and meta-regression analysis. JAMA. 2002;287(8):1022–8.
37. Kalsi T, Babic-Illman G, Ross PJ, Maisey NR, Hughes S, Fields P, et al. The impact of comprehensive geriatric assessment interventions on tolerance to chemotherapy in older people. Br J Cancer. 2015;112(9):1435–44.
38. Lin JS, Whitlock EP, Eckstrom E, Fu R, Perdue LA, Beil TL, et al. Challenges in synthesizing and interpreting the evidence from a systematic review of multifactorial interventions to prevent functional decline in older adults. J Am Geriatr Soc. 2012;60(11):2157–66.
39. Muder RR. Pneumonia in residents of long-term care facilities: epidemiology, etiology, management, and prevention. Am J Med. 1998;105(4):319–30.
40. Marrie TJ. Pneumonia in the long-term-care facility. Infect Control Hosp Epidemiol. 2002;23(3):159–64.
41. Leng S, Chaves P, Koenig K, Walston J. Serum interleukin-6 and hemoglobin as physiological correlates in the geriatric syndrome of frailty: a pilot study. J Am Geriatr Soc. 2002;50(7):1268–71.
42. Leng SX, Cappola AR, Andersen RE, Blackman MR, Koenig K, Blair M, et al. Serum levels of insulin-like growth factor-I (IGF-I) and dehydroepiandrosterone sulfate (DHEA-S), and their relationships with serum interleukin-6, in the geriatric syndrome of frailty. Aging Clin Exp Res. 2004;16(2):153–7.
43. Walston J, McBurnie MA, Newman A, Tracy RP, Kop WJ, Hirsch CH, et al. Frailty and activation of the inflammation and coagulation systems with and without clinical comorbidities: results from the cardiovascular health study. Arch Intern Med. 2002;162(20):2333–41.
44. Schoufour JD, Echteld MA, Bastiaanse LP, Evenhuis HM. The use of a frailty index to predict adverse health outcomes (falls, fractures, hospitalization, medication use, comorbid conditions) in people with intellectual disabilities. Res Dev Disabil. 2015;38:39–47.
45. Desquilbet L, Margolick JB, Fried LP, Phair JP, Jamieson BD, Holloway M, et al. Relationship between a frailty-related phenotype and progressive deterioration of the immune system in HIV-infected men. J Acquir Immune Defic Syndr. 2009;50(3):299–306.
46. Franceschi C, Bonafe M. Centenarians as a model for healthy aging. Biochem Soc Trans. 2003;31(2):457–61.
47. Garagnani P, Giuliani C, Pirazzini C, Olivieri F, Bacalini MG, Ostan R, et al. Centenarians as super-controls to assess the biological relevance of genetic risk factors for common age-related diseases: a proof of principle on type 2 diabetes. Aging (Albany, NY). 2013;5(5):373–85.
48. Cruz-Jentoft AJ, Baeyens JP, Bauer JM, Boirie Y, Cederholm T, Landi F, et al. Sarcopenia: European consensus on definition and diagnosis: report of the European Working Group on sarcopenia in older people. Age Ageing. 2010;39(4):412–23.
49. Rothman MD, Leo-Summers L, Gill TM. Prognostic significance of potential frailty criteria. J Am Geriatr Soc. 2008;56(12):2211–6.

50. Romanick M, Thompson LV, Brown-Borg HM. Murine models of atrophy, cachexia, and sarcopenia in skeletal muscle. Biochim Biophys Acta. 2013;1832(9):1410–20.
51. Walston J, Fedarko N, Yang H, Leng S, Beamer B, Espinoza S, et al. The physical and biological characterization of a frail mouse model. J Gerontol A Biol Sci Med Sci. 2008;63(4):391–8.
52. Whitehead JC, Hildebrand BA, Sun M, Rockwood MR, Rose RA, Rockwood K, et al. A clinical frailty index in aging mice: comparisons with frailty index data in humans. J Gerontol A Biol Sci Med Sci. 2014;69(6):621–32.
53. Schoufour JD, Echteld MA, Boonstra A, Groothuismink ZM, Evenhuis HM. Biochemical measures and frailty in people with intellectual disabilities. Age Ageing. 2016;45(1):142–8.
54. Harris TB, Ferrucci L, Tracy RP, Corti MC, Wacholder S, Ettinger Jr WH, et al. Associations of elevated interleukin-6 and C-reactive protein levels with mortality in the elderly. Am J Med. 1999;106(5):506–12.
55. Ferrucci L, Harris TB, Guralnik JM, Tracy RP, Corti MC, Cohen HJ, et al. Serum IL-6 level and the development of disability in older persons. J Am Geriatr Soc. 1999;47(6):639–46.
56. Cohen HJ, Pieper CF, Harris T, Rao KM, Currie MS. The association of plasma IL-6 levels with functional disability in community-dwelling elderly. J Gerontol A Biol Sci Med Sci. 1997;52(4):M201–8.
57. Leng SX, Yang H, Walston JD. Decreased cell proliferation and altered cytokine production in frail older adults. Aging Clin Exp Res. 2004;16(3):249–52.
58. Csordas A, Fuchs D, Frangieh AH, Reibnegger G, Stähli BE, Cahenzly M, et al. Immunological markers of frailty predict outcomes beyond current risk scores in aortic stenosis following transcatheter aortic valve replacement: Role of neopterin and tryptophan. IJC Metab Endocr. 2016;10:7–15.
59. Visser M, Pahor M, Taaffe DR, Goodpaster BH, Simonsick EM, Newman AB, et al. Relationship of interleukin-6 and tumor necrosis factor-alpha with muscle mass and muscle strength in elderly men and women: the Health ABC Study. J Gerontol A Biol Sci Med Sci. 2002;57(5):M326–32.
60. Taaffe DR, Harris TB, Ferrucci L, Rowe J, Seeman TE. Cross-sectional and prospective relationships of interleukin-6 and C-reactive protein with physical performance in elderly persons: MacArthur studies of successful aging. J Gerontol A Biol Sci Med Sci. 2000;55(12):M709–15.
61. Payette H, Roubenoff R, Jacques PF, Dinarello CA, Wilson PW, Abad LW, et al. Insulin-like growth factor-1 and interleukin 6 predict sarcopenia in very old community-living men and women: the Framingham Heart Study. J Am Geriatr Soc. 2003;51(9):1237–43.
62. Schaap LA, Pluijm SM, Deeg DJ, Visser M. Inflammatory markers and loss of muscle mass (sarcopenia) and strength. Am J Med. 2006;119(6):526.e9–17.
63. Greiwe JS, Cheng B, Rubin DC, Yarasheski KE, Semenkovich CF. Resistance exercise decreases skeletal muscle tumor necrosis factor alpha in frail elderly humans. FASEB J. 2001;15(2):475–82.
64. Grunfeld C, Zhao C, Fuller J, Pollack A, Moser A, Friedman J, et al. Endotoxin and cytokines induce expression of leptin, the ob gene product, in hamsters. J Clin Invest. 1996;97(9):2152–7.
65. Sarraf P, Frederich RC, Turner EM, Ma G, Jaskowiak NT, Rivet 3rd DJ, et al. Multiple cytokines and acute inflammation raise mouse leptin levels: potential role in inflammatory anorexia. J Exp Med. 1997;185(1):171–5.
66. Leng SX, Xue QL, Tian J, Walston JD, Fried LP. Inflammation and frailty in older women. J Am Geriatr Soc. 2007;55(6):864–71.
67. Leng SX, Xue Q-L, Tian J, Huang Y, Yeh S-H, Fried LP. Associations of neutrophil and monocyte counts with frailty in community-dwelling disabled older women: results from the Women's Health and Aging Studies I. Exp Gerontol. 2009;44(8):511–6.
68. Collerton J, Martin-Ruiz C, Davies K, Hilkens CM, Isaacs J, Kolenda C, et al. Frailty and the role of inflammation, immunosenescence and cellular ageing in the very old: cross-sectional findings from the Newcastle 85+ Study. Mech Ageing Dev. 2012;133(6):456–66.

69. Fernandez-Garrido J, Navarro-Martinez R, Buigues-Gonzalez C, Martinez-Martinez M, Ruiz-Ros V, Cauli O. The value of neutrophil and lymphocyte count in frail older women. Exp Gerontol. 2014;54:35–41.

70. Baylis D, Bartlett DB, Syddall HE, Ntani G, Gale CR, Cooper C, et al. Immune-endocrine biomarkers as predictors of frailty and mortality: a 10-year longitudinal study in community-dwelling older people. Age (Dordr). 2013;35(3):963–71.

71. Qu T, Walston JD, Yang H, Fedarko NS, Xue QL, Beamer BA, et al. Upregulated ex vivo expression of stress-responsive inflammatory pathway genes by LPS-challenged CD14(+) monocytes in frail older adults. Mech Ageing Dev. 2009;130(3):161–6.

72. Yao X, Hamilton RG, Weng NP, Xue QL, Bream JH, Li H, et al. Frailty is associated with impairment of vaccine-induced antibody response and increase in post-vaccination influenza infection in community-dwelling older adults. Vaccine. 2011;29(31):5015–21.

73. Desquilbet L, Jacobson LP, Fried LP, Phair JP, Jamieson BD, Holloway M, et al. HIV-1 infection is associated with an earlier occurrence of a phenotype related to frailty. J Gerontol A Biol Sci Med Sci. 2007;62(11):1279–86.

74. Ng TP, Camous X, Nyunt MSZ, Vasudev A, Tan CTY, Feng L, et al. Markers of T-cell senescence and physical frailty: insights from Singapore Longitudinal Ageing Studies. NPJ Aging Mech Dis. 2015;1:15005.

75. Wikby A, Maxson P, Olsson J, Johansson B, Ferguson FG. Changes in CD8 and CD4 lymphocyte subsets, T cell proliferation responses and non-survival in the very old: the Swedish longitudinal OCTO-immune study. Mech Ageing Dev. 1998;102(2-3):187–98.

76. Semba RD, Margolick JB, Leng S, Walston J, Ricks MO, Fried LP. T cell subsets and mortality in older community-dwelling women. Exp Gerontol. 2005;40(1-2):81–7.

77. Schmaltz HN, Fried LP, Xue QL, Walston J, Leng SX, Semba RD. Chronic cytomegalovirus infection and inflammation are associated with prevalent frailty in community-dwelling older women. J Am Geriatr Soc. 2005;53(5):747–54.

78. Wang GC, Kao WH, Murakami P, Xue QL, Chiou RB, Detrick B, et al. Cytomegalovirus infection and the risk of mortality and frailty in older women: a prospective observational cohort study. Am J Epidemiol. 2010;171(10):1144–52.

79. Gibson KL, Wu YC, Barnett Y, Duggan O, Vaughan R, Kondeatis E, et al. B-Cell diversity decreases in old age and is correlated with poor health status. Aging Cell. 2009;8(1):18–25.

80. Chou CH, Hwang CL, Wu YT. Effect of exercise on physical function, daily living activities, and quality of life in the frail older adults: a meta-analysis. Arch Phys Med Rehabil. 2012;93(2):237–44.

81. Theou O, Stathokostas L, Roland KP, Jakobi JM, Patterson C, Vandervoort AA, et al. The effectiveness of exercise interventions for the management of frailty: a systematic review. J Aging Res. 2011;2011:569194.

82. Chin AP, van Uffelen JGZ, Riphagen IR, van Mechelen W. The functional effects of physical training in frail older people: a systematic review. Sports Med. 2008;38(9):781–93.

83. Lee PHLY, Chan D-C. Interventions targeting geriatric frailty: a systematic review. J Gerontol Geriatr. 2012;3:47–52.

84. Gine-Garriga M, Roque-Figuls M, Coll-Planas L, Sitja-Rabert M, Salva A. Physical exercise interventions for improving performance-based measures of physical function in community-dwelling, frail older adults: a systematic review and meta-analysis. Arch Phys Med Rehabil. 2014;95(4):753–69. e3.

85. Peterson MJ, Sloane R, Cohen HJ, Crowley GM, Pieper CF, Morey MC. Effect of telephone exercise counseling on frailty in older veterans: project LIFE. Am J Mens Health. 2007;1(4):326–34.

86. Bartali B, Frongillo EA, Bandinelli S, Lauretani F, Semba RD, Fried LP, et al. Low nutrient intake is an essential component of frailty in older persons. J Gerontol A Biol Sci Med Sci. 2006;61(6):589–93.

87. Beasley JM, LaCroix AZ, Neuhouser ML, Huang Y, Tinker L, Woods N, et al. Protein intake and incident frailty in the Women's Health Initiative observational study. J Am Geriatr Soc. 2010;58(6):1063–71.

88. Bollwein J, Diekmann R, Kaiser MJ, Bauer JM, Uter W, Sieber CC, et al. Distribution but not amount of protein intake is associated with frailty: a cross-sectional investigation in the region of Nurnberg. Nutr J. 2013;12:109.

89. Kim CO, Lee KR. Preventive effect of protein-energy supplementation on the functional decline of frail older adults with low socioeconomic status: a community-based randomized controlled study. J Gerontol A Biol Sci Med Sci. 2013;68(3):309–16.

90. Payette H, Boutier V, Coulombe C, Gray-Donald K. Benefits of nutritional supplementation in free-living, frail, undernourished elderly people: a prospective randomized community trial. J Am Diet Assoc. 2002;102(8):1088–95.

91. Chin APM, de Jong N, Schouton EG, Hiddink GJ, Koh FJ. Physical exercise and/or enriched foods for functional improvement in frail, independently living elderly: a randomized control trial. Arch Phys Med Rehabil. 2001;82(6):811–7.

92. Ng TP, Feng L, Nyunt MS, Feng L, Niti M, Tan BY, et al. Nutritional, physical, cognitive, and combination interventions and frailty reversal among older adults: a randomized controlled trial. Am J Med. 2015;128(11):1225–36. e1.

93. Tieland M, Dirks ML, van der Zwaluw N, Verdijk LB, van de Rest O, de Groot LC, et al. Protein supplementation increases muscle mass gain during prolonged resistance-type exercise training in frail elderly people: a randomized, double-blind, placebo-controlled trial. J Am Med Dir Assoc. 2012;13(8):713–9.

94. Chan DC, Tsou HH, Yang RS, Tsauo JY, Chen CY, Hsiung CA, et al. A pilot randomized controlled trial to improve geriatric frailty. BMC Geriatr. 2012;12:58.

95. Cesari M, Vellas B, Hsu FC, Newman AB, Doss H, King AC, et al. A physical activity intervention to treat the frailty syndrome in older persons-results from the LIFE-P study. J Gerontol A Biol Sci Med Sci. 2015;70(2):216–22.

96. Bauer J, Biolo G, Cederholm T, Cesari M, Cruz-Jentoft AJ, Morley JE, et al. Evidence-based recommendations for optimal dietary protein intake in older people: a position paper from the PROT-AGE Study Group. J Am Med Dir Assoc. 2013;14(8):542–59.

97. Gray SL, LaCroix AZ, Aragaki AK, McDermott M, Cochrane BB, Kooperberg CL, et al. Angiotensin-converting enzyme inhibitor use and incident frailty in women aged 65 and older: prospective findings from the Women's Health Initiative Observational Study. J Am Geriatr Soc. 2009;57(2):297–303.

98. Sumukadas D, Witham MD, Struthers AD, McMurdo ME. Effect of perindopril on physical function in elderly people with functional impairment: a randomized controlled trial. CMAJ. 2007;177(8):867–74.

99. Ensrud KE, Ewing SK, Fredman L, Hochberg MC, Cauley JA, Hillier TA, et al. Circulating 25-hydroxyvitamin D levels and frailty status in older women. J Clin Endocrinol Metab. 2010;95(12):5266–73.

100. Ensrud KE, Blackwell TL, Cauley JA, Cummings SR, Barrett-Connor E, Dam TT, et al. Circulating 25-hydroxyvitamin D levels and frailty in older men: the osteoporotic fractures in men study. J Am Geriatr Soc. 2011;59(1):101–6.

101. Dhesi JK, Jackson SH, Bearne LM, Moniz C, Hurley MV, Swift CG, et al. Vitamin D supplementation improves neuromuscular function in older people who fall. Age Ageing. 2004;33(6):589–95.

102. Bischoff-Ferrari HA, Dawson-Hughes B, Orav EJ, Staehelin HB, Meyer OW, Theiler R, et al. Monthly high-dose vitamin D treatment for the prevention of functional decline: a randomized clinical trial. JAMA Intern Med. 2016;176(2):175–83.

103. Snyder PJ, Peachey H, Hannoush P, Berlin JA, Loh L, Lenrow DA, et al. Effect of testosterone treatment on body composition and muscle strength in men over 65 years of age. J Clin Endocrinol Metab. 1999;84(8):2647–53.

104. Kenny AM, Kleppinger A, Annis K, Rathier M, Browner B, Judge JO, et al. Effects of transdermal testosterone on bone and muscle in older men with low bioavailable testosterone levels, low bone mass, and physical frailty. J Am Geriatr Soc. 2010;58(6):1134–43.

105. Page ST, Amory JK, Bowman FD, Anawalt BD, Matsumoto AM, Bremner WJ, et al. Exogenous testosterone (T) alone or with finasteride increases physical performance, grip strength, and lean body mass in older men with low serum T. J Clin Endocrinol Metab. 2005;90(3):1502–10.
106. Snyder PJ, Bhasin S, Cunningham GR, Matsumoto AM, Stephens-Shields AJ, Cauley JA, et al. Effects of testosterone treatment in older men. N Engl J Med. 2016;374(7):611–24.
107. Baylis D, Bartlett DB, Syddall HE, Ntani G, Gale CR, Cooper C, et al. Immune-endocrine biomarkers as predictors of frailty and mortality: a 10-year longitudinal study in community-dwelling older people. Age. 2013;35(3):963–71.
108. Taekema DG, Ling CH, Blauw GJ, Meskers CG, Westendorp RG, de Craen AJ, et al. Circulating levels of IGF1 are associated with muscle strength in middle-aged- and oldest-old women. Eur J Endocrinol. 2011;164(2):189–96.
109. Liu H, Bravata DM, Olkin I, Nayak S, Roberts B, Garber AM, et al. Systematic review: the safety and efficacy of growth hormone in the healthy elderly. Ann Intern Med. 2007;146(2): 104–15.
110. Criscione LG, St Clair EW. Tumor necrosis factor-alpha antagonists for the treatment of rheumatic diseases. Curr Opin Rheumatol. 2002;14(3):204–11.
111. Lahaye C, Tatar Z, Dubost JJ, Soubrier M. Overview of biologic treatments in the elderly. Joint Bone Spine. 2015;82(3):154–60.
112. Fairhall N, Kurrle SE, Sherrington C, Lord SR, Lockwood K, John B, et al. Effectiveness of a multifactorial intervention on preventing development of frailty in pre-frail older people: study protocol for a randomised controlled trial. BMJ Open. 2015;5(2):007091.
113. Fairhall N, Aggar C, Kurrle SE, Sherrington C, Lord S, Lockwood K, et al. Frailty Intervention Trial (FIT). BMC Geriatr. 2008;8:27.
114. Cameron ID, Fairhall N, Langron C, Lockwood K, Monaghan N, Aggar C, et al. A multifactorial interdisciplinary intervention reduces frailty in older people: randomized trial. BMC Med. 2013;11(1):65.

Lifestyle Interventions to Improve Immunesenescence

10

David B. Bartlett and Kim M. Huffman

Abstract

Regular participation in exercise and physical activity is associated with many health benefits including improvements in metabolic and cardiorespiratory function. However, as we get older the time and intensity at which we exercise is severely reduced. Physical inactivity now accounts for a considerable proportion of age-related disease and mortality. These diseases are associated with an increased systemic inflammatory milieu and immunesenescence. Regular exercise and physical activity has been suggested to exert anti-inflammatory and anti-immunesenescence effects, potentially delaying the health declines with ageing. No immune cells are impervious to the effects of ageing and exercise appears to modify many immunological functions. Regular exercise has been shown to improve neutrophil microbicidal functions which reduce the risk of infectious disease. Exercise participation is also associated with increased immune cell telomere length, and may be related to improved vaccine responses. The anti-inflammatory effect of regular exercise and negative energy balance is evident by reduced inflammatory immune cell signatures and lower inflammatory cytokine concentrations. In this chapter we will discuss the role of physical activity and energy balance in modifying the immune system. Specifically we discuss what role each plays on limiting the incidence of immunesenescence in older adults.

Keywords

Innate immunity • Adaptive immunity • Immunesenescence • Inflammation • Exercise • Physical activity • Metabolism

D.B. Bartlett, Ph.D. (✉)
Duke Molecular Physiology Unit, Duke School of Medicine, Duke University, Durham, NC, USA

Duke Molecular Physiology Institute, Department of Medicine, Duke University Medical Center, Durham, NC 27701, USA
e-mail: david.bartlett@dm.duke.edu

K.M. Huffman, M.A., Ph.D.
Duke Molecular Physiology Unit, Duke School of Medicine, Duke University, Durham, NC, USA

© Springer International Publishing Switzerland 2017
V. Bueno et al. (eds.), *The Ageing Immune System and Health*,
DOI 10.1007/978-3-319-43365-3_10

10.1 Introduction

Poor lifestyle choices associated with advancing age are characterized by physical inactivity and increased fat mass. It is estimated that physical inactivity is associated with around 9% of premature deaths worldwide (5.3 million) from non-communicable diseases [1]. The burden of chronic disease associated with physical inactivity ranges from 6% for coronary heart disease to 10% for breast and colon cancer [1]. The aetiology of many of these diseases is associated with immunesenescence and inflammation [2]. This chapter will discuss studies which have assessed the effects of physical activity, structured exercise and energy balance on immune function and inflammation in the aged, and what still needs to be done to better understand the benefits of exercise for immunity. In light of the majority of research on exercise immunology being conducted in younger individuals or animal models, these will be referred to in the absence of data pertaining to older humans.

Maintaining regular habitual physical activity levels throughout life is associated with a number of health benefits including, reduced risk of cardiovascular disease (CVD), diabetes and stroke; as well as reduced physical disability and mortality [3]. However, after the age of 45 years there is a progressive and sharp decline in both time and intensity of physical activity which is attributed to a lack of time, motivation and knowledge of the health benefits [4]. Reduced physical activity is associated with a positive energy balance leading to increased adiposity and subsequently systemic inflammation [5]. The link between immunesenescence and inflammation is clear with the two having major negative effects on health [6]. In the following sections we will describe the effects of different measures of exercise on immune function and the role exercise has in modifying immunesenescence and inflammation.

10.2 Innate Immune Response to Exercise and Physical Activity

10.2.1 Neutrophils

As described in Chap. 1, ageing is characterized by reduced neutrophil phagocytic capacity, chemotaxis, ROS production and extracellular trap formation. However, despite the well characterized dysfunctions with age, less is known about the effects of exercise on neutrophil function in the aged human.

Elevated neutrophil counts accompany increased inflammation with age and the increased ratio of neutrophils to lymphocytes is associated with many age-related diseases including cancer [7]. Compared to more active individuals, less active and overweight individuals have higher circulating neutrophil counts [8]. Six weeks of cycling exercise, 1–6 times per week at around 80–90% of aerobic capacity (VO_{2max}) was associated with reduced neutrophil counts in women over 50 years old [8]. Furthermore, long-term exercise training of around 80 minutes per day, 5 days per week in those over 70 years resulted in lower neutrophil counts than age-matched non-exercising individuals [9]. Although these studies suggest an intervention and years of exercise training can reduce neutrophil numbers the data are not consistent [10, 11].

More likely is that neutrophil numbers, with no obvious disease state, are homeostatically maintained in relation to body mass. In overweight individuals, aerobic exercise training can reduce neutrophil numbers, which is associated with glucose homeostasis and body fat content [8, 12]. These data suggest that neutrophils are susceptible to inflammatory perturbations associated with the metabolic consequences of a positive energy balance and that exercise training can impact neutrophils by improving energy balance and glucose homeostasis [13]. Indeed, poor neutrophil function is associated with increased fasting glucose levels [14]. Thus, neutrophil numbers might be reduced in those with underlying inflammatory or metabolic disparities and may explain differences in healthy older individuals [11, 15].

Although neutrophil numbers are associated with poor disease prognosis, their function is critical to their response. Neutrophil function is a difficult mechanism to assess in large scale studies due to their short half-life and inherent activation capacity. This is further augmented by the acute effects of exercise, which promotes a transient increase in cell numbers following acute exercise and the delay before returning to basal levels. Thus exercise studies are complicated by when and how to assess neutrophil function and to-date it remains unclear the ideal time that should be used.

Subsequently, few studies have assessed neutrophil function in response to exercise or physical activity. When compared to age-matched individuals completing 5000 steps/day, physical activity of over 10,000 steps/day was associated with better neutrophil chemotaxis towards the inflammatory chemokine IL-8 [11]. No differences were observed for phagocytic capacity or reactive oxygen species (ROS) production towards ingested *E. coli*, which is in contrast to other reports [10, 16]. In these studies, increased physical activity in older individuals was associated with better phagocytic capacity than age matched physically inactive elders [10, 16]. In younger individuals, neutrophil chemotaxis, mitochondrial function and phagocytosis have all been improved either by an acute exercise bout or months of aerobic training [17, 18]. However, ROS production has been shown to be impaired during periods of excessive exercise training [19].

In older people with disease, the role of exercise on neutrophil function is even more complicated. Interpretation of results is difficult because it is unclear their pathogenic role. For example, in respiratory diseases such as COPD neutrophil function may be elevated or reduced and still contribute to disease pathology [20]. Twelve weeks of moderate-intensity walking was recently shown to reduce basal levels of ROS production and increase CD62L expression, suggesting an improved inflammatory and immune-surveillance phenotype in older individuals [21]. However, aquatic exercise for 8-months was associated with no improvements in phagocytic or chemotactic properties of neutrophils in women with fibromyalgia [22].

The release of extracellular DNA by neutrophils (NETs) as a mechanism of infection and tissue damage control has received significant attention recently. Although cell free DNA is associated with a number of inflammatory and disease conditions, NET production is reduced in healthy elders [23]. As NET production is associated with ROS activity and ROS activity is also reduced this would seem causally linked. However, elevated and uncontrolled NET production is associated with contributing to cell free DNA and thus disease such as atherosclerosis [24].

An interesting study by Beiter and colleagues suggested that NET release was increased following acute exercise [25]. They suggest that this response in a sterile, non-inflammatory state promotes immune homeostasis and protection from chronic inflammation through cross-talk with other immune cells such as dendritic and T-cells [25].

Taken together, exercise and physical activity has the potential to modify neutrophil cell composition and function in older adults. More work is required to fully understand the effects of exercise on neutrophil functions such as chemotaxis and microbicidal activity in seniors and whether there is relationship with overall improved health outcomes. Furthermore, little is known about the intensity, duration and type of exercise which can provide benefits to neutrophil function. Considering neutrophil function is severely impaired with age and is associated with the pathogenesis of many age-associated diseases it is perplexing that so little is known.

10.2.2 Monocytes, Macrophages and Dendritic Cells

Monocytes make up a relatively small proportion of circulating leukocytes (2–12 %) but due to their diverse role in immune function and inflammation have received more interest in the exercise literature than neutrophils. With age, there is a tendency for monocytes to adopt a pro-inflammatory phenotype as characterized by increased expression of CD16 and synthesis of TNFα, IL-6 and IL-1β in the basal state [26, 27]. However, they and their differentiated forms, macrophages and dendritic cells, are less functional as characterized by reduced TLR stimulation, cytokine production and antigen presentation upon stimulation [28, 29]. Together, dysfunctional monocytes are associated with the pathogenesis of inflammatory diseases such as CVD and peripheral artery disease [30]. Exercise may serve as a tool to reduce the inflammatory nature of monocytes and macrophages [31].

In response to acute short bouts of exercise such as running and cycling, peripheral monocyte numbers are increased. The pro-inflammatory CD16+ population accounts for the largest percentage change and is likely indicative of a surveillance mechanism [32]. Although following acute exercise the numbers and percentages of subtype's return to basal levels there is a marked reduction in CD16+ monocytes after exercise training interventions [32, 33]. It is unclear whether the repeated exercise sessions of training is removing CD16+ monocytes from the circulation or whether the differentiation to a CD16+ monocyte is perturbed in relation to changes in inflammation, metabolism and endocrine mediators. Timmerman and colleagues showed that older active individuals had lower percentages of CD16+ monocytes than inactive age and sex matched individuals [33]. Following 12-weeks of aerobic exercise at 70–80 % of heart rate reserve combined with resistance exercise the inactive individuals reduced the total percentage of CD16+ monocytes [33]. Recently, resistance training with no weight loss was sufficient to reduce the CD16+ population [34]. Thus, both aerobic and resistance training have been shown to influence CD16+ monocyte numbers, highlighting these cells as potential exercise-mediated inflammatory biomarkers of disease risk.

In addition to reduced CD16 expression, 12-weeks of combined aerobic and resistance training was sufficient to reduce basal and LPS stimulated TNFα production [33]. Although TLR-4 expression was unaffected by training and the inflammatory response likely due to the reduced percentage of CD16+ cells, others have shown lowered TLR-4 and TLR-2 expression with training in the elderly [31, 35, 36]. Following prolonged acute exercise TLR expression is reduced for several hours. Mechanisms behind the exercise induced reduction in TLR expression are unknown however it has been suggested that anti-inflammatory cytokines and/or glucocorticoid production may have an impact [37]. Additionally, 12-weeks of a combined resistance and aerobic exercise intervention resulted in an increase in expression of the co-stimulatory molecule CD80 [38]. Reduced expression of CD80 on monocytes has been shown to relate to poor vaccine responses in the old due to reduced activation potential of T-cells and antigen presentation capabilities [39]. Phagocytosis of pathogens is reduced with ageing however the effects of exercise have received little attention. To date the only study to assess phagocytosis was in middle aged men and showed no changes following 12 weeks of training [40].

Migration of peripheral blood monocytes and their subsequent differentiation to macrophages in adipose tissue is critical to the development of sustained inflammation during obesity. Whether exercise can moderate this process and reduce inflammation is unknown. Although it is unclear if exercise or the negative energy balance associated with exercise reduces adipocyte mass, the production of chemokines (MIP1α and MCP-1) responsible for macrophage infiltration is reduced with exercise [41]. Vascular adhesion molecules such as ICAM-1 are increased with ageing and chronic disease but have been shown to be reduced by 6 months of aerobic exercise [42]. Therefore it is plausible that exercise reduces the migratory capacity of monocytes into adipose tissue [43]. Macrophages with the inflammatory phenotype (M1) are preferentially associated with increased adipose tissue [44]. The effects of exercise on macrophage phenotype in humans remains to be determined, however in mice it has been recently suggested that exercise induces a phenotypic switch of M1 to anti-inflammatory/regulatory M2 macrophages in the adipose tissue as well as reducing M1 infiltration [43].

Few studies have assessed the relationship of exercise and dendritic cell (DC) function. Similar to monocytes, DCs with age exhibit an increased pro-inflammatory phenotype and reduced antigen presenting capacity [28, 45]. Recently, a single bout of acute exercise was sufficient to increase numbers of monocyte derived DCs without compromising function [46]. Therefore, in those with compromised DC function it is plausible that exercise can facilitate generation of more DCs and at least maintain DC function. In individuals with non-small cell lung cancer, 8-weeks of tai chi classes was sufficient to increase numbers of myeloid DCs in the blood, potentially providing enhanced immunity against cancer [47]. Both myeloid and plasmacytoid DC's are mobilized by acute exercise with myeloid DC's increasing to a greater extent [48]. Given the importance of DC function in relation to health and disease risk, more research is needed to determine the role of exercise on DC function.

Taken together, it is clear that exercise and physical activity have an impact on monocytes, macrophages and dendritic cells. Although obtaining tissue resident macrophages and DCs is a laborious, expensive and highly invasive task in relation to exercise, the generation of these cells from exercise modified blood monocytes will provide greatly needed insight to the functions of such critical immunologic and inflammatory mediators.

10.2.3 NK Cells

Natural Killer (NK) cells are of significant interest in relation to ageing, health and exercise, whereby their functions are severely impaired by age but with exercise and physical activity functions are often improved [49]. NK cells are predominantly associated with recognition and elimination of transformed malignant cells. However, their role has expanded to include modulation of adaptive immune responses, anti-microbial activity and resolution of inflammation [50]. Therefore, reduced NK cell function may play a pivotal role in the development of malignancy and increased viral infections seen in the old.

The majority of NK cell and exercise research has been conducted in younger individuals, with few studies assessing age or disease in relation to exercise. Increased NK cell cytotoxicity (NKCC) against the tumor cell line K562 was evident in response to 6 months of moderate intensity (50–65 % VO_{2max}) aerobic training in old individuals [15]. However, no differences for NKCC were observed for 12 months of moderate-intensity (40–75 % VO_{2max}) aerobic training in post-menopausal women [51]. No differences in NKCC or NK cell numbers were evident following 12 weeks of walking exercises in older sedentary women, although a highly active age-matched group had better NKCC than the sedentary women [52]. No differences were observed between physically active and inactive men, although total numbers of NK cells were higher in active men [10, 53]. These results are unexpected given the relationship with increased physical fitness and cancer survivorship [54]. NK cells are highly responsive and mobile cells, rapidly mobilized during periods of homeostatic imbalance. Therefore, it is plausible that NK-cell function is not being assessed to its fullest potential. For example, baseline NK cell function may not give the full picture and instead an acute bout of exercise may be required to assess global NK cell function.

NK-cells consist of multiple sub-types which appear to mirror that of the T-cell arm, with distinct cytotoxic and regulatory properties [55]. NK cells are highly responsive to acute exercise, being mobilized from the marginal to the peripheral compartments rapidly [56]. Bigley and colleagues have recently improved the potential for assessing NK cell function using acute bouts of exercise to mobilize NK-cells, giving a more comprehensive picture of functional capacity [57, 58]. Acute exercise is associated with a redistribution of NK cells with altered phenotypes and results in increased NKCC towards a number of HLA-deficient tumor cell lines [58]. A recent study by Pedersen and colleagues showed that 30 days of wheel running in mice was associated with an exercise dependent mobilization of NK cells which reduced tumor burden in 5 cancer models [59]. Unfortunately neither of these studies was conducted in older humans or animals. Until these studies are conducted in old animals or humans it remains unclear whether exercise and physical activity can override the effects of NK cell dysfunction in the old.

10.3   Adaptive Immune Function Responses to Exercise and Physical Activity

10.3.1   T Cells

The T-cell compartment with ageing is associated with significant changes; especially in respect to the composition of subsets. In the older adult, the T-cell compartment is characterized by an altered CD4:CD8 ratio which is dominated by increased proportions of late-stage differentiated CD8+ T-cells (CD28-, CD57+, KLRG1+, CD27-) [60]. These cells are often specific to epitopes of latent viral infections such as cytomegalovirus (CMV) and Epstein Barr Virus (EBV). Subsequently, production of IL-2 is lower and their proliferative capacity is reduced, while increases are observed for pro-inflammatory cytokines [61]. Together, this characterization is termed the "Immune Risk Profile" (IRP) and is associated with increased risk of mortality in the older adult [62].

A considerable number of studies have assessed the effects of acute and chronic exercise on measures of T-cell immunesenescence including T cell subsets, phenotype, proliferation, cytokine production, chemotaxis, and co-stimulatory capacity. At least one comprehensive review is published each year [63–67]. We will attempt to condense this large body of research, focusing primarily on exercise studies specifically in the old.

T-cells are mobilized into the peripheral blood compartment by an acute bout of exercise in both young and older individuals [53, 68]. This mobilization is characterized by a preferential increase in late-stage differentiated T-cells [68, 69]. Following exercise the balance and numbers of cells are restored over the space of 1–24 h. Where these cells come from, where they go and what the impact this has on T-cell homeostasis or immune function remains unknown. However it has been proposed that dysfunctional cells are selectively eliminated allowing for improved T-cell function [70]. To confirm this, exercise training in the old should elicit improved T-cell function and altered subset phenotypes when compared to non-exercise or sedentary elders.

A recent cross-sectional analysis of over 100 men, aged 18–61 years suggested that increased exercise capacity, as measured by VO_{2max}, was associated with reduced late-stage differentiated T-cells [71]. While findings did not withstand statistical adjustment for VO_{2max} they offer an interesting insight into T-cell repertoire response to physical activity with age. In support of this, 6 months of aerobic exercise in older sedentary men and women resulted in increased proliferative capacity and a reduced memory phenotype of T-cells [15]. Furthermore, 12 weeks of aerobic and resistance training increased the numbers of CD28 expressing CD8+ cells, suggesting an enhanced immune capacity [38]. Although responses to vaccinations appear improved following aerobic exercise training, it is unclear whether this is improved T-cell function, DC function or both [72, 73]. Unfortunately these findings are not consistent in the majority of studies. Aerobic and/or resistance training ranging from 12 weeks to 12 months have shown no changes in proliferative capacity [51, 52, 74], naïve to memory phenotype [15], T-cell markers of immunesenescence [75], CD4:CD8 ratios, regulatory T-cell numbers or γδT-cell numbers [65].

Cell telomere length is at present a widely accepted measure of biological and immunological ageing in humans. Telomeres are the repetitive DNA nucleotide caps present on the ends of each chromosome, protecting the chromosome from fusion with other chromosomes. During each cellular division, erosion of the DNA cap protecting the end of chromosomes results in shorter telomeres and over time will result in reduce telomere length with age. Cross-sectional studies have highlighted that compared to inactive older individuals, physically active older individuals have longer telomere lengths [76, 77]. Increased telomere lengths in young individuals peripheral blood mononuclear cells (PBMCs) following acute exercise have also been reported [68]. Recently, Silva and colleagues suggested that older exercisers had increased T-cell telomere lengths when compared to age and BMI matched controls [77]. Interestingly, these longer telomere lengths were found to be associated with CD8+ T-cells lacking expression of CD28, suggesting exercise as a means to attenuate the effects of immune cell ageing. In support of this, telomere lengths are also longer in less frail older individuals suggesting that regular activity and exercise training has a significant impact of cellular health [78, 79].

10.4 Exercise and Chronic Low-Grade Inflammation

A common characteristic of ageing is an increased chronic low-grade systemic inflammation which contributes to a number of diseases including diabetes, cardiovascular disease, Alzheimer's and cancer [80, 81]. Termed 'inflammageing' it is characterized by increased concentrations of pro-inflammatory cytokines and reduced concentrations of anti-inflammatory cytokines [81]. Although the cause of inflammation is multifactorial, immunesenescence and obesity are thought to contribute significantly [82]. Both ageing and a positive energy balance are associated with metabolic and physiological changes leading to increased adiposity. The result is an increased production of cytokines leading to catabolic effects in muscle and deprives tissue of nutrients through altering insulin signaling leading to chronic damage [83]. Couple this with a dysfunctional immune system prone to producing increased inflammatory cytokines and the effects on health are potentially great [6, 78, 84]. Moreover, the anti-inflammatory effect of exercise is associated with a reduced risk of chronic metabolic and cardiorespiratory diseases [66]. Three possible mechanisms have been suggested: the reduction in visceral fat mass, synthesis and secretion of anti-inflammatory cytokines from skeletal muscle and modification of immunesenescence (discussed previously) [31, 66, 85].

This anti-inflammatory effect of exercise may be mediated by the release of the typically pro-inflammatory cytokine IL-6 from working muscle [86]. IL-6 is a pleiotropic cytokine that signals differently depending on its origins and target cells [85]. Subsequently, during exercise there is a transient increase in muscle derived IL-6 which promotes the synthesis of anti-inflammatory cytokines such as IL-10 and IL-1 receptor agonist (RA) from immune cells [87]. Immune cell infiltration into adipose tissue is also reduced by exercise. Subsequently, reduced adipocyte infiltration by neutrophils, macrophages and perhaps T-cells has been shown with exercise [13, 43]. Furthermore, exercise promotes increases in the adrenal hormones cortisol and adrenaline which have anti-inflammatory effects on immune cell production of TNFα and IL-1β [88–90].

Cross-sectional studies are often used to determine associations with physical activity and systemic inflammation. It remains unclear whether exercise alone reduces fat mass, however with exercise there is a shift in energy balance which is associated with reduced visceral fat mass [91]. Critically, the metabolism of adipocytes is altered and they become sensitive to insulin and glucose metabolism which is associated with increased adiponectin, a metabolic hormone responsible for fatty acid metabolism [92]. Therefore, reduced inflammatory cytokines such as CRP, IL-6 and TNFα coincide with elevated concentrations of adiponectin in physically active older individuals [93–96]. Similarly, exercise interventions ranging from 12 weeks to 6 months of aerobic and/or resistance exercise in older individuals resulted in reduced markers of inflammation and increased anti-inflammatory markers [97, 98].

Taken together exercise appears to promote an anti-inflammatory response which is mediated by altered adipocyte function and improved energy metabolism leading to suppression of pro-inflammatory cytokine production in immune cells. Although others have shown no effects of exercise and physical activity on inflammation [11, 99, 100], these studies are atypical and the majority report exercise as anti-inflammatory.

10.5   Potential Mechanisms

To-date it remains unclear the mechanisms associated with exercise mediated immune enhancement, but it is likely to be multi-factorial. Exercise and energy balance have direct influences on a number of systems with immunomodulatory properties, including metabolism, inflammation, endocrine, oxidative stress and muscle function [66]. Many of these systems are exclusively associated with adipose tissue and adipose influenced inflammation. Even in the absence of weight changes with exercise training, fitness and metabolism are often improved and go hand in hand with multi-organ functional improvements. Unlike model in vitro experimental designs, it is difficult to untangle these interactions in human research.

Simpson and colleagues have suggested that repeated bouts of acute exercise are sufficient to mobilize cells from the marginal space into the peripheral blood whereby dysfunctional T-cells are selectively removed [65, 70]. Although others have suggested that apoptosis is increased following exercise it remains unclear which cell sub-types are being 'selectively' removed [101, 102]. If exercise does selectively remove dysfunctional T-cells then reconstitution of T-cells with functional perhaps immature or naïve T-cells would imply an improved thymic function. Thymic atrophy is a hallmark of ageing and is associated with reduced naïve T-cell output [103]. Recently, administration of IL-7 therapy is associated with a temporary increase in thymic mass and production of naïve T-cells [103]. Interestingly exercise induces production of IL-7 from the muscle and IL-7 has been observed to increase systemically following exercise, suggesting a possible feedback mechanism [104]. Measures of thymic function in response to exercise are limited with few studies assessing recent thymic emigrants (RTE), which can be assessed by enumerating T cells bearing T-cell receptor excision circles (TREC). Recently it was shown that compared to age-matched controls elite athletes had lower TREC levels [105]. This would imply a negative impact for older exercisers in response to naïve T-cell output.

The impact of exercise and physical activity on adaptive immune function in older adults appears less pronounced than effects on the innate arm. Neutrophils and monocytes are relatively short-lived cells and at any given time the peripheral blood contains unknown proportions of cells at the end of their lifespan than at the beginning. It is plausible that exercise also selectively removes more innate cells at the end of their lifespan, improving the balance and function. Neutrophils at the end of their lifespan begin to express markers for homeostatic clearance including CXCR4 and lose markers such as CD62L, and are more susceptible to apoptotic signaling [106]. Acute exercise mobilizes neutrophils into the blood which are more susceptible to apoptosis [18]. Additionally, regular exercise training is associated with reduced apoptosis and increased expression of CD62L on neutrophils; suggesting that the balance of 'old' to 'newer' neutrophils has been changed [18, 21]. Further work needs to be conducted to determine if this is the case and understand the mechanisms. Neutrophils mature in the bone marrow; therefore it is likely that the effects of exercise may be influencing bone marrow biology.

Recently, gene and microRNA analysis has attempted to determine changes in immune cell functional mechanisms responding to acute exercise. Radom-Aizik and colleagues are leading the way with changes in response to exercise in multiple cell types [107–110]. These studies employ analysis of gene changes in response to acute exercise, and although interpretation is difficult due to the compositional changes in immune cells caused by acute exercise the results are fascinating. A short bout of acute exercise was sufficient to alter gene expression profiles of monocytes that were characterized by anti-inflammatory and anti-atherogenic alterations. These included downregulation of *TNFα*, *TLR4* and *CD36* genes and upregulation of *epiregulin* (*EREG*) and *CXCR4* genes [107]. These findings were recently strengthened by Abbasi and colleagues who determined gene profiles following high intensity acute exercise [111]. Similarly, 986 NK-cell gene alterations were characterized by increased tumor cytotoxicity such as tumor-cell communications, p53 signaling pathway, tumor cell adherens and focal adhesion genes [108]. Both neutrophils and PBMCs appear to have altered gene expression characterized by increased inflammatory and growth and repair processes [109, 110]. The next step for these studies is to assess the effect of an exercise intervention and determine whether gene expression results in protein translation. Together, these studies suggest that although we may be unable to see phenotype and functional changes of immune cells, the potential is there for them to adapt to exercise training.

10.6   Considerations and Future Perspectives

As with all human studies, interpreting results with caution is required. The majority of cited literature assesses immune responses in healthy individuals and thus their immune system may be working optimally in preventing infection and disease.

Therefore, although at times no effects for exercise are observed this should be expected. Subsequently, when exercise studies are applied to individuals with chronic diseases such as cancer or CVD the results are often positive [9, 112, 113]. Because these individuals are immunocompromised by their disease, exercise often promotes enhancements in function.

We have specifically avoided discussing the interactions of exercise and immune function in those infected with latent viral infections for the following reasons. Immunesenescence is associated with repeated antigenic exposure over time and an exhaustion of immune function [114]. Latent infections with herpes viruses such as CMV and EBV are known to promote clonal expansion of viral specific T-cells. These cells occupy a relatively large immune space compared to uninfected hosts and may contribute to a reduced capacity to deal with novel pathogens. Indeed, in the IRP the presence of CMV and EBV is associated with mortality in the elderly [115]. These viral infections are characterized by an altered phenotype of T-and NK-cells and acute bouts of exercise are known to mobilize cell phenotypes differently in infected versus uninfected [116]. However, the role of CMV and EBV in the pathogenesis of ageing is not clear and its relationship with inflammation and physical function with age is questionable [117, 118]. Thus, the suggested need to remove these viral specific cells via exercise may not be the mechanism by which exercise improves measures of senescence. In fact removal of these cells may increase re-activation of CMV, EBV or their cousin herpes virus varicella-zoster, responsible for shingles in the elderly. Furthermore, CMV specific T-cells have reduced expression of the programmed death-1 receptor (PD-1) [119]. Malignant tumors often express the PD-1 ligand which inhibits T-cell tumor cytotoxicity; therefore removal of CMV specific T-cells may increase the risk of cancer.

As the immune system of the older adult is dominated by the actions of the myeloid compared to the lymphoid lineage, much more work is needed to determine the role of exercise in modifying viral specific T-cells. Clearly, there will be a number of such cells which are senescent and contribute to poor immunity; however this remains to be determined.

A major limitation of the role exercise and energy balance plays on immune function in the old is the lack of standardized testing protocols, participant populations and low participant numbers. A number of studies have compared interventions of various exercise intensities, modes and durations on health measures such as glucose tolerance, cardiovascular function and physical function [99, 120]. With the recent development of the National Institute of Health common fund designed to understand the molecular transducers of exercise perhaps we will have a clearer understanding of how exercise can modify the immune system.

In summary, exercise training and regular physical activity likely promotes an anti-inflammatory and anti-immunesenescence effect. A repeated acute bout of activity increases heart rate and works multiple muscle groups stimulating a mobilization of immune cells into the peripheral blood. Where these cells go and what happens after exercise remains unknown, but this repetitive exercise over time can modify aspects of immunesenescence. Whether these effects are rejuvenating or delay the impact of inflammation and senescence remains to be seen. What is clear is that physical activity and energy balance influence the immune and inflammatory responses reducing the risk of age-associated disease and infection.

References

1. Lee I-M et al. Effect of physical inactivity on major non-communicable diseases worldwide: an analysis of burden of disease and life expectancy. Lancet. 2012;380(9838):219–29.
2. Pawelec G. Immunity and ageing in man. Exp Gerontol. 2006;41(12):1239–42.
3. Agahi N, Parker MG. Leisure activities and mortality. J Aging Health. 2008;20(7):855–71.
4. Caspersen CJ, Pereira MA, Curran KM. Changes in physical activity patterns in the United States, by sex and cross-sectional age. Med Sci Sports Exerc. 2000;32(9):1601–9.
5. Fantuzzi G. Adipose tissue, adipokines, and inflammation. J Allergy Clin Immunol. 2005;115(5):911–9. quiz 920.
6. Baylis D et al. Understanding how we age: insights into inflammaging. Longevity Healthspan. 2013;2(1):8.
7. Polat M et al. Neutrophil to lymphocyte and platelet to lymphocyte ratios increase in ovarian tumors in the presence of frank stromal invasion. Clin Transl Oncol. 2015;18(5):457–63.
8. Michishita R et al. Effect of exercise therapy on monocyte and neutrophil counts in overweight women. Am J Med Sci. 2010;339(2):152–6.
9. Moro-Garcia MA et al. Frequent participation in high volume exercise throughout life is associated with a more differentiated adaptive immune response. Brain Behav Immun. 2014;39:61–74.
10. Yan H et al. Effect of moderate exercise on immune senescence in men. Eur J Appl Physiol. 2001;86(2):105–11.
11. Bartlett DB et al. Habitual physical activity is associated with the maintenance of neutrophil migratory dynamics in healthy older adults. Brain Behav Immun. 2016;56:12–20.
12. Johannsen NM et al. Effect of different doses of aerobic exercise on total white blood cell (WBC) and WBC subfraction number in postmenopausal women: results from DREW. PLoS One. 2012;7(2), e31319.
13. Kawanishi N et al. Exercise training attenuates neutrophil infiltration and elastase expression in adipose tissue of high-fat-diet-induced obese mice. Physiol Rep. 2015;3(9):pii:e12534.
14. Saito Y et al. The influence of blood glucose on neutrophil function in individuals without diabetes. Luminescence. 2013;28(4):569–73.
15. Woods JA et al. Effects of 6 months of moderate aerobic exercise training on immune function in the elderly. Mech Ageing Dev. 1999;109(1):1–19.
16. Sasaki S et al. Effects of regular exercise on neutrophil functions, oxidative stress parameters and antibody responses against 4-hydroxy-2-nonenal adducts in middle aged humans. Exerc Immunol Rev. 2013;19:60–71.
17. Syu GD, Chen HI, Jen CJ. Differential effects of acute and chronic exercise on human neutrophil functions. Med Sci Sports Exerc. 2012;44(6):1021–7.
18. Syu GD, Chen HI, Jen CJ. Severe exercise and exercise training exert opposite effects on human neutrophil apoptosis via altering the redox status. PLoS One. 2011;6(9), e24385.
19. Pyne MDB. Regulation of neutrophil function during exercise. Sports Med. 1994;17(4):245–58.
20. Stockley JA et al. Aberrant neutrophil functions in stable chronic obstructive pulmonary disease: the neutrophil as an immunotherapeutic target. Int Immunopharmacol. 2013;17(4):1211–7.
21. Takahashi M et al. Low-volume exercise training attenuates oxidative stress and neutrophils activation in older adults. Eur J Appl Physiol. 2013;113(5):1117–26.
22. Bote ME et al. An exploratory study of the effect of regular aquatic exercise on the function of neutrophils from women with fibromyalgia: Role of IL-8 and noradrenaline. Brain Behav Immun. 2014;39:107–12.
23. Hazeldine J et al. Impaired neutrophil extracellular trap formation: a novel defect in the innate immune system of aged individuals. Aging Cell. 2014;13(4):690–8.
24. Megens RT et al. Presence of luminal neutrophil extracellular traps in atherosclerosis. Thromb Haemost. 2012;107(3):597–8.
25. Beiter T et al. Neutrophils release extracellular DNA traps in response to exercise. J Appl Physiol. 2014;117(3):325–33.

26. Belge K et al. The proinflammatory CD14+CD16+DR++ monocytes are a major source of TNF. J Immunol. 2002;168(7):3536–42.
27. Ziegler-Heitbrock L. The CD14+ CD16+ blood monocytes: their role in infection and inflammation. J Leukoc Biol. 2007;81(3):584–92.
28. Panda A et al. Age-associated decrease in TLR function in primary human dendritic cells predicts influenza vaccine response. J Immunol. 2010;184(5):2518–27.
29. Shaw A et al. Aging of the innate immune system. Curr Opin Immunol. 2010;22(4):507–13.
30. Pande RL et al. Association of monocyte TNFα expression and serum inflammatory biomarkers with walking impairment in PAD. J Vasc Surg. 2015;61(1):155–61.
31. Flynn MG, McFarlin BK. Toll-like receptor 4: link to the anti-inflammatory effects of exercise? Exerc Sport Sci Rev. 2006;34:176–81.
32. Simpson RJ et al. Toll-like receptor expression on classic and pro-inflammatory blood monocytes after acute exercise in humans. Brain Behav Immun. 2009;23(2):232–9.
33. Timmerman KL et al. Exercise training-induced lowering of inflammatory (CD14+CD16+) monocytes: a role in the anti-inflammatory influence of exercise? J Leukoc Biol. 2008;84(5):1271–8.
34. Markofski MM et al. Resistance exercise training-induced decrease in circulating inflammatory CD14+CD16+ monocyte percentage without weight loss in older adults. Eur J Appl Physiol. 2014;114(8):1737–48.
35. Stewart LK et al. Influence of exercise training and age on CD14+ cell-surface expression of toll-like receptor 2 and 4. Brain Behav Immun. 2005;19(5):389–97.
36. McFarlin BK et al. Physical activity status, but not age, influences inflammatory biomarkers and toll-like receptor 4. J Gerontol A Biol Sci Med Sci. 2006;61(4):388–93.
37. Lancaster GI et al. The physiological regulation of toll-like receptor expression and function in humans. J Physiol. 2005;563(3):945–55.
38. Shimizu K et al. Monocyte and T-cell responses to exercise training in elderly subjects. J Strength Cond Res. 2011;25(9):2565–72.
39. van Duin D, Shaw AC. Toll-like receptors in older adults. J Am Geriatr Soc. 2007;55(9):1438–44.
40. Schaun MI et al. The effects of periodized concurrent and aerobic training on oxidative stress parameters, endothelial function and immune response in sedentary male individuals of middle age. Cell Biochem Funct. 2011;29(7):534–42.
41. Gano LB et al. Increased proinflammatory and oxidant gene expression in circulating mononuclear cells in older adults: amelioration by habitual exercise. Physiol Genomics. 2011;43(14):895–902.
42. Zoppini G et al. Effects of moderate-intensity exercise training on plasma biomarkers of inflammation and endothelial dysfunction in older patients with type 2 diabetes. Nutr Metab Cardiovasc Dis. 2006;16(8):543–9.
43. Kawanishi N et al. Exercise training inhibits inflammation in adipose tissue via both suppression of macrophage infiltration and acceleration of phenotypic switching from M1 to M2 macrophages in high-fat-diet-induced obese mice. Exerc Immunol Rev. 2010;16:105–18.
44. Woods JA, Davis JM. Exercise, monocyte/macrophage function, and cancer. Med Sci Sports Exerc. 1994;26(2):147–56.
45. Agrawal A et al. Altered innate immune functioning of dendritic cells in elderly humans: a role of phosphoinositide 3-kinase-signaling pathway. J Immunol. 2007;178(11):6912–22.
46. LaVoy ECP et al. A single bout of dynamic exercise by healthy adults enhances the generation of monocyte-derived-dendritic cells. Cell Immunol. 2015;295(1):52–9.
47. Liu J et al. Effect of Tai Chi on mononuclear cell functions in patients with non-small cell lung cancer. BMC Complement Altern Med. 2015;15:3.
48. Suchanek O et al. Intensive physical activity increases peripheral blood dendritic cells. Cell Immunol. 2010;266(1):40–5.
49. Bigley AB et al. Can exercise-related improvements in immunity influence cancer prevention and prognosis in the elderly? Maturitas. 2013;76(1):51–6.

50. Hazeldine J, Lord JM. The impact of ageing on natural killer cell function and potential consequences for health in older adults. Ageing Res Rev. 2013;12(4):1069–78.
51. Campbell PT et al. Effect of exercise on in vitro immune function: a 12-month randomized, controlled trial among postmenopausal women. J Appl Physiol (1985). 2008;104(6):1648–55.
52. Nieman DC et al. Physical activity and immune function in elderly women. Med Sci Sports Exerc. 1993;25(7):823–31.
53. Shinkai S et al. Physical activity and immune senescence in men. Med Sci Sports Exerc. 1995;27(11):1516–26.
54. Jones LW et al. Cardiorespiratory exercise testing in clinical oncology research: systematic review and practice recommendations. Lancet Oncol. 2008;9(8):757–65.
55. Sun JC, Lanier LL. NK cell development, homeostasis and function: parallels with CD8+ T cells. Nat Rev Immunol. 2011;11(10):645–57.
56. Shephard RJ, Shek PN. Effects of exercise and training on natural killer cell counts and cytolytic activity: a meta-analysis. Sports Med. 1999;28(3):177–95.
57. Bigley AB et al. NK-cells have an impaired response to acute exercise and a lower expression of the inhibitory receptors KLRG1 and CD158a in humans with latent cytomegalovirus infection. Brain Behav Immun. 2012;26(1):177–86.
58. Bigley AB et al. Acute exercise preferentially redeploys NK-cells with a highly-differentiated phenotype and augments cytotoxicity against lymphoma and multiple myeloma target cells. Brain Behav Immun. 2014;39:160–71.
59. Pedersen L et al. Voluntary running suppresses tumor growth through epinephrine- and IL-6-dependent NK cell mobilization and redistribution. Cell Metab. 2016;23(3):554–62.
60. Ouyang Q et al. Age-associated accumulation of CMV-specific CD8+ T cells expressing the inhibitory killer cell lectin-like receptor G1 (KLRG1). Exp Gerontol. 2003;38(8):911–20.
61. Bauer ME, Fuente M. The role of oxidative and inflammatory stress and persistent viral infections in immunosenescence. Mech Ageing Dev. 2016;pii:S0047-6374(16)30001-X.
62. Derhovanessian E, Larbi A, Pawelec G. Biomarkers of human immunosenescence: impact of Cytomegalovirus infection. Curr Opin Immunol. 2009;21(4):440–5.
63. Turner JE. Is immunosenescence influenced by our lifetime "dose" of exercise? Biogerontology. 2016;17(3):581–602.
64. Kohut ML, Senchina DS. Reversing age-associated immunosenescence via exercise. Exerc Immunol Rev. 2004;10:6–41.
65. Simpson RJ et al. Exercise and the aging immune system. Ageing Res Rev. 2012.
66. Gleeson M et al. The anti-inflammatory effects of exercise: mechanisms and implications for the prevention and treatment of disease. Nat Rev Immunol. 2011;11(9):607–15.
67. Walsh NP et al. Position statement part one: immune function and exercise. Exerc Immunol Rev. 2011;17:6–63.
68. Simpson RJ et al. Senescent phenotypes and telomere lengths of peripheral blood T-cells mobilized by acute exercise in humans. Exerc Immunol Rev. 2010;16:40–55.
69. Simpson RJ et al. High-intensity exercise elicits the mobilization of senescent T lymphocytes into the peripheral blood compartment in human subjects. J Appl Physiol. 2007;103(1):396–401.
70. Simpson RJ. Aging, persistent viral infections, and immunosenescence: can exercise "make space"? Exerc Sport Sci Rev. 2011;39(1):23–33.
71. Spielmann G et al. Aerobic fitness is associated with lower proportions of senescent blood T-cells in man. Brain Behav Immun. 2011;25(8):1521–9.
72. Kohut ML et al. Moderate exercise improves antibody response to influenza immunization in older adults. Vaccine. 2004;22(17-18):2298–306.
73. Pascoe AR, Fiatarone Singh MA, Edwards KM. The effects of exercise on vaccination responses: a review of chronic and acute exercise interventions in humans. Brain Behav Immun. 2014;39:33–41.
74. Flynn MG et al. Effects of resistance training on selected indexes of immune function in elderly women. J Appl Physiol. 1999;86(6):1905–13.

75. Kapasi ZF et al. Effects of an exercise intervention on immunologic parameters in frail elderly nursing home residents. J Gerontol A Biol Sci Med Sci. 2003;58(7):636–43.
76. Cherkas LF et al. The association between physical activity in leisure time and leukocyte telomere length. Arch Intern Med. 2008;168(2):154–8.
77. Silva LC et al. Moderate and intense exercise lifestyles attenuate the effects of aging on telomere length and the survival and composition of T cell subpopulations. Age (Dordr). 2016;38(1):24.
78. Baylis D et al. Inflammation, telomere length, and grip strength: a 10-year longitudinal study. Calcif Tissue Int. 2014;95(1):54–63.
79. Woo J et al. Telomere length is associated with decline in grip strength in older persons aged 65 years and over. Age (Dordr). 2014;36(5):9711.
80. De Martinis M et al. Inflammation markers predicting frailty and mortality in the elderly. Exp Mol Pathol. 2006;80(3):219–27.
81. Franceschi C et al. Inflammaging and anti-inflammaging: a systemic perspective on aging and longevity emerged from studies in humans. Mech Ageing Dev. 2007;128:92–105.
82. De Martinis M et al. Inflamm-ageing and lifelong antigenic load as major determinants of ageing rate and longevity. FEBS Lett. 2005;579(10):2035–9.
83. Pedersen M et al. Circulating levels of TNF-alpha and IL-6-relation to truncal fat mass and muscle mass in healthy elderly individuals and in patients with type-2 diabetes. Mech Ageing Dev. 2003;124(4):495–502.
84. Baylis D et al. Immune-endocrine biomarkers as predictors of frailty and mortality: a 10-year longitudinal study in community-dwelling older people. Age (Dordr). 2013;35(3):963–71.
85. Petersen AMW, Pedersen BK. The anti-inflammatory effect of exercise. J Appl Physiol. 2005;98(4):1154–62.
86. Pedersen BK, Edward F. Adolph distinguished lecture: muscle as an endocrine organ: IL-6 and other myokines. J Appl Physiol. 2009;107(4):1006–14.
87. Steensberg A et al. IL-6 enhances plasma IL-1ra, IL-10, and cortisol in humans. Am J Physiol Endocrinol Metab. 2003;285(2):E433–7.
88. Galbo H. Hormonal and metabolic adaptation to exercise. New York, NY: Thieme-Stratton Inc; 1983. p. 1–116.
89. Cupps TR, Fauci AS. Corticosteroid-mediated immunoregulation in man. Immunol Rev. 1982;65:133–55.
90. Bergmann M. Attenuation of catecholamine-induced immunosuppression in whole blood from patients with sepsis. Shock. 1999;12:421–7.
91. Ross R, Bradshaw AJ. The future of obesity reduction: beyond weight loss. Nat Rev Endocrinol. 2009;5(6):319–25.
92. Mujumdar PP et al. Long-term, progressive, aerobic training increases adiponectin in middle-aged, overweight, untrained males and females. Scand J Clin Lab Invest. 2011;71:101–7.
93. Colbert LH et al. Physical activity, exercise, and inflammatory markers in older adults: findings from the health, aging and body composition study. J Am Geriatr Soc. 2004;52(7):1098–104.
94. Kullo IJ, Khaleghi M, Hensrud DD. Markers of inflammation are inversely associated with VO2 max in asymptomatic men. J Appl Physiol (1985). 2007;102(4):1374–9.
95. Pedersen BK, Bruunsgaard H. Possible beneficial role of exercise in modulating low-grade inflammation in the elderly. Scand J Med Sci Sports. 2003;13(1):56–62.
96. Heaney JL et al. Serum free light chains are reduced in endurance trained older adults: Evidence that exercise training may reduce basal inflammation in older adults. Exp Gerontol. 2016;77:69–75.
97. Nicklas BJ et al. Exercise training and plasma C-reactive protein and interleukin-6 in elderly people. J Am Geriatr Soc. 2008;56(11):2045–52.
98. Nunes PR et al. Effect of resistance training on muscular strength and indicators of abdominal adiposity, metabolic risk, and inflammation in postmenopausal women: controlled and randomized clinical trial of efficacy of training volume. Age (Dordr). 2016;38(2):40.

99. Church TS et al. Exercise without weight loss does not reduce C-reactive protein: the INFLAME study. Med Sci Sports Exerc. 2010;42(4):708–16.
100. Huffman KM et al. Response of high-sensitivity C-reactive protein to exercise training in an at-risk population. Am Heart J. 2006;152(4):793–800.
101. Kruger K et al. Exercise affects tissue lymphocyte apoptosis via redox-sensitive and Fas-dependent signaling pathways. Am J Physiol Regul Integr Comp Physiol. 2009;296(5):R1518–27.
102. Kruger K, Mooren FC. T cell homing and exercise. Exerc Immunol Rev. 2007;13:37–54.
103. Aspinall R, Mitchell W. Reversal of age-associated thymic atrophy: treatments, delivery, and side effects. Exp Gerontol. 2008;43(7):700–5.
104. Haugen F et al. IL-7 is expressed and secreted by human skeletal muscle cells. Am J Physiol Cell Physiol. 2010;298(4):C807–16.
105. Prieto-Hinojosa A et al. Reduced thymic output in elite athletes. Brain Behav Immun. 2014;39:75–9.
106. Zhang D et al. Neutrophil ageing is regulated by the microbiome. Nature. 2015;525(7570):528–32.
107. Radom-Aizik S et al. Impact of brief exercise on circulating monocyte gene and microRNA expression: implications for atherosclerotic vascular disease. Brain Behav Immun. 2014;39:121–9.
108. Radom-Aizik S et al. Impact of brief exercise on peripheral blood NK cell gene and microRNA expression in young adults. J Appl Physiol (1985). 2013;114(5):628–36.
109. Radom-Aizik S et al. Effects of exercise on microRNA expression in young males peripheral blood mononuclear cells. Clin Transl Sci. 2012;5(1):32–8.
110. Radom-Aizik S et al. Effects of 30 min of aerobic exercise on gene expression in human neutrophils. J Appl Physiol. 2008;104(1):236–43.
111. Abbasi A et al. Exhaustive exercise modifies different gene expression profiles and pathways in LPS-stimulated and un-stimulated whole blood cultures. Brain Behav Immun. 2014;39:130–41.
112. Glass OK et al. Effect of aerobic training on the host systemic milieu in patients with solid tumours: an exploratory correlative study. Br J Cancer. 2015;112(5):825–31.
113. Sallam N, Laher I. Exercise modulates oxidative stress and inflammation in aging and cardio-vascular diseases. Oxid Med Cell Longev. 2016;2016:7239639.
114. Pawelec G. Hallmarks of human "immunosenescence": adaptation or dysregulation? Immun Ageing. 2012;9(1):15.
115. Pawelec G, Ferguson FG, Wikby A. The SENIEUR protocol after 16 years. Mech Ageing Dev. 2001;122(2):132–4.
116. Spielmann G et al. The effects of age and latent cytomegalovirus infection on the redeployment of CD8+ T cell subsets in response to acute exercise in humans. Brain Behav Immun. 2014;39:142–51.
117. Bartlett DB et al. The age-related increase in low-grade systemic inflammation (Inflammaging) is not driven by cytomegalovirus infection. Aging Cell. 2012;11(5):912–5.
118. Goldeck D et al. No strong correlations between serum cytokine levels, CMV serostatus and hand-grip strength in older subjects in the Berlin BASE-II cohort. Biogerontology. 2015;17(1):189–98.
119. Trautmann L et al. Upregulation of PD-1 expression on HIV-specific CD8+ T cells leads to reversible immune dysfunction. Nat Med. 2006;12(10):1198–202.
120. Kraus WE. Effects of the amount and intensity of exercise on plasma lipoproteins. N Engl J Med. 2002;347:1483–92.

Index

Printed in the United States
By Bookmasters